Mechanical Self-Assembly

Xi Chen
Editor

Mechanical Self-Assembly

Science and Applications

Springer

Editor
Xi Chen
Department of Earth and Environmental Engineering
Columbia University
New York, NY, USA

ISBN 978-1-4614-4561-6 ISBN 978-1-4614-4562-3 (eBook)
DOI 10.1007/978-1-4614-4562-3
Springer New York Heidelberg Dordrecht London

Library of Congress Control Number: 2012952936

Printed on acid-free paper

Springer is part of Springer Science+Business Media (www.springer.com)

Preface

The traditional micro/nanofabrication techniques based on photolithography are intrinsically expensive, time-consuming, effective to only a limited set of photoresists, and not directly applicable to nonplanar surfaces. Among the conceptually new strategies that offer possible routes to both small features and low costs, self-assembly involving thin sheets, films, and multilayers has become the new cornerstone. In self-assembly, small-scale objects spontaneously organize and aggregate into stable, well-defined structures, which are guided by the characteristics (e.g. surface, electrical, and mechanical functionalities) of the subunits, and the final structures are reached by equilibrating to the form of the lowest free energy. The design of components that organize themselves into desired patterns and functions is the key to applications of self-assembly. In addition, self-assembly can be used with soft lithography and/or controlled deposition to transfer and create complex features in various small material systems.

Typical molecular self-assembly involves noncovalent interactions, such as electrostatic, van der Waals, and hydrogen bonds. To assemble larger components beyond the molecular level, physical forces including magnetic, capillary, dispersion, and entropic interactions are also involved. Presented in this book is a novel category of self-assembly driven by mechanical forces: Mechanical Self-Assembly, where controlled mechanical buckling is shown to be able to self-assemble ordered patterns in thin films.

Traditionally, buckling and buckle delamination are among the most frequently encountered failure modes in thin films. While a significant amount of efforts have been put together to understand the failure mechanisms and to improve the reliability of thin film systems, through mechanical self-assembly, they can become useful and underpin a new fabrication technique. From a practical point of view, mechanical self-assembly is complementary to, and sometimes offers advantages over, the conventional physical and chemical self-assembly processes. From a fundamental point of view, it also provides a rich opportunity to extend our understanding of the mechanics in thin films.

Mechanical self-assembly is arguably one of the cheapest, quickest, and simplest techniques for manufacturing and patterning structures at the micron and submicron scales; it works with most common thin film materials, offers various selections of patterns, and directly applicable to planar and nonplanar surfaces. The subject of mechanical self-assembly is a wide open area with numerous exciting potentials in engineering and biology remain to be explored.

The book collects a number of recent advances in the area of mechanical self-assembly, including fundamental wrinkling and delamination mechanics, experimental techniques, applications to fabrication and morphogenesis. Chapter 1 presents an overview of mechanical self-assembly in nature and engineering. Chapter 2 discusses the spontaneous patterns formed by growth of biological systems. Theory of shaping by growth is given in Chap. 3. Chapter 4 illustrates experimental thin film wrinkle morphologies. Relevant theoretical background is given in Chap. 5. In Chap. 6, the surface crease phenomena are discussed. Chapters 7 and 8 present delamination buckling/channels self-assembled in thin films. Finally, Chap. 9 extends spontaneous buckles on curved substrates.

New York, NY, USA Xi Chen

Contents

Contributors

Xi Chen Department of Earth and Environmental Engineering, Columbia Nanomechanics Research Center, Columbia University New York, NY, USA

SV Lab, International Center of Applied Mechanics, School of Aerospace, Xi'an Jiaotong University Xi'an, China

Department of Civil & Environmental Engineering, Hanyang University Seoul, Korea

Fangliang Gao Department of Aerospace Engineering, Iowa State University Ames, IA, USA

Wei Hong Department of Aerospace Engineering, Iowa State University Ames, IA, USA

Rui Huang Department of Aerospace Engineering and Engineering Mechanics, University of Texas at Austin Austin, TX, USA

Chansoo Kim Future Convergence Research Division, Korea Institute of Science and Technology (KIST) Seoul, Republic of Korea

Myoung-Woon Moon Future Convergence Research Division, Korea Institute of Science and Technology (KIST) Seoul, Republic of Korea

Eran Sharon The Racah Institute of Physics, The Hebrew University of Jerusalem Jerusalem, Israel

Ashkan Vaziri Department of Mechanical and Industrial Engineering, Northeastern University Boston, MA, USA

Alex A. Volinsky Department Mechanical Engineering, University of South Florida Tampa, FL, USA

Patrick Waters Department Mechanical Engineering, University of South Florida Tampa, FL, USA

Jie Yin Department of Earth and Environmental Engineering, Columbia Nanomechanics Research Center, Columbia University New York, NY, USA

Chapter 1
Mechanical Self-Assembly in Nature

Xi Chen and Jie Yin

Abstract Mother Nature provides unlimited inspirations of ordered patterns across vast scales: from the helical DNA and lipid bilayers at the submicron level, to the skin and tissue wrinkle at the millimeter level, to the ordered shapes in plants and animals at the meter level, and to the geological features at the mega scale. Many of these intriguing patterns are underpinned by mechanics-driven processes, including spontaneous buckling deformation.

1.1 Patterns and Morphologies in Nature

Over thousands of years of evolution, Mother Nature is able to self-assemble precise, differentiable, and intriguing morphologies on all scales from nano- to macro-size. For example, in cells, through bending and stretching deformation, the planar bilayer with phospholipids in solution can self-assemble into the liposome and micelles as shown in Fig. 1.1a. The self-folding of polypeptide chains into coiled proteins through bending and twisting is another famous molecular self-assembly example shown in Fig. 1.1b. At the tissue level, taking the development morphology of human's brain cortex for example, unlike a smooth kidney or

X. Chen (✉)
Department of Earth and Environmental Engineering, Columbia Nanomechanics Research Center, Columbia University, New York, NY 10027, USA

SV Lab, International Center of Applied Mechanics, School of Aerospace, Xi'an Jiaotong University, Xi'an 710049, China

Department of Civil & Environmental Engineering, Hanyang University, Seoul 133-791, Korea
e-mail: xichen@columbia.edu

J. Yin
Department of Earth and Environmental Engineering, Columbia Nanomechanics Research Center, Columbia University, New York, NY 10027, USA

X. Chen (ed.), *Mechanical Self-Assembly: Science and Applications*,
DOI 10.1007/978-1-4614-4562-3_1, © Springer Science+Business Media New York 2013

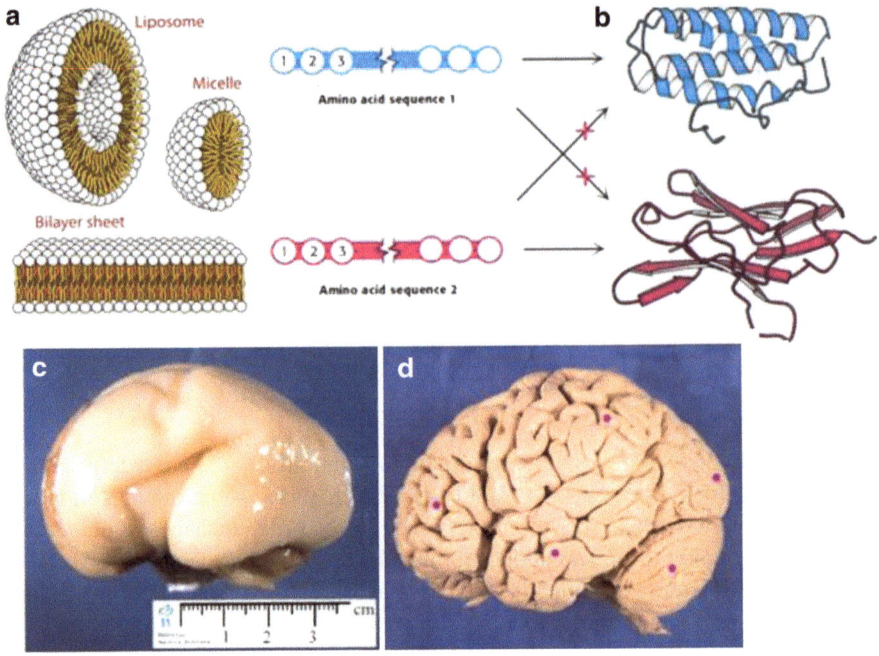

Fig. 1.1 Self-assembly in cells and tissues: (**a**) Three typical self-assembled structures of lipid bilayers in solution: liposome, micelle, and bilayer sheet. (**b**) Self-folding of polypeptide chains into proteins. (**c**) The smooth brain cortex of a fetus at 22 weeks. (**d**) The corrugated morphology of an adult human being's brain cortex. Yin [1], reprinted with permission

spleen, cerebral cortex is full of wrinkles and folds (see Fig. 1.1d). However, in the fetus period it is smooth (see Fig. 1.1c). At the early stage, as neurons continue to divide, grow, and migrate, the cortex folds and self-assembles a recognizable but unique pattern of bumps and grooves.

In the world of plants and fruits, Mother Nature creates more varieties of patterns and morphologies. Taking the different and intricate shapes found in leaves for example, leaves are seldom observed in perfect planar shapes; indeed, one of the most common morphologies is the saddle-like global appearance as shown in Fig. 1.2a (e.g., the rhododendron leaf). Moreover, a pear leaf often possesses a local wavy morphology along its edge (see Fig. 1.2b). Such a locally rippled edge is often found in the pond weed leaves; in addition, they often take the global curling and twisting topologies (see Fig. 1.2c).

A very special morphology is the tubular leaf shape found in rhododendron leaves in winter (see Fig. 1.2d), where the normally open and nearly planar leaf self-folds into a nearly closed tubular one as the temperature decreases. Other shapes including the doubly curled (see Fig. 1.2e) and spiral morphology (see Fig. 1.2f) are also found in leaves and tendrils. Figure 1.2g shows that most flowers have the similar wavy margins as leaves. A more complicated pattern is also observed in the sunflowers' heads and pine cones, where they have the beautiful and precise Fibonacci number patterns (see Fig. 1.2h, i).

Fig. 1.2 Different morphologies and patterns of leaves and flowers observed in nature. (**a**) Saddle-like rhododendron leaf. (**b**) A pear leaf with local wavy edge. (**c**) The global curled and twisted morphology of pond weed leaves. (**d**) The tubular morphology of rhododendron leaf (same as (**a**) except that the present one is in winter). (**e**) Doubly curled leaf. (**f**) Spiral morphology of a leaf. (**g**) Flowers with wavy margins in growth. Fibonacci number patterns in (**h**) sunflowers' head and (**i**) pine cones. Yin [1], reprinted with permission

Unlike balloons with a smooth surface, some fruits and vegetables distinguish themselves from others by possessing undulating surface morphologies as shown in Fig. 1.3.

For example, the small pumpkins such as acorn squashes show ten equidistant ridges running from the stem to the tip. Like the development of brain cortex, a pumpkin experiences a morphology transition from a smooth at newborn to the 10-ridged shape beyond a critical moment (and the overall ridged shape remains stable with continuous growth), as shown in Fig. 1.4. Similar 10-ridged morphology can also be found in the Korean melon (golden melon) and ridged gourd (or silk gourd, *Luffa acutanglula*). For another breed of big pumpkins, it often has about 20 ridges. Striped cavern tomatoes and bell peppers have 4–6 ribs to characterize their unique

Fig. 1.3 Examples of several natural fruits that exhibit distinctive buckle-like undulations as their global appearances. The number on the right corner shows the number of ridges. Chen and Yin [16], reproduced by permission of The Royal Society of Chemistry. The article in which this figure was originally published is located at the following link: http://pubs.rsc.org/en/content/articlelanding/2010/sm/c0sm00401d

Fig. 1.4 Morphology transition during the growth of a typical pumpkin from smooth surface to undulating morphologies. Yin [1], reprinted with permission

appearances. Different from the ribbed pattern, reticular cantaloupes demonstrate a netted pattern on their surface. In all these examples, the fruits/vegetables possess relatively stiff skin and relatively compliant flesh, and the overall shape is approximately spheroidal. Another special case is the wax apples, whose overall shape is conical and they exhibit a beautiful skin wrinkle-like ridged morphology on their surfaces. Note that for the same breed of fruits, the ridge number remains essentially unchanged despite their different sizes. Interestingly, similar ridged patterns can be also found in butterfly, bollworm and tobacco budworm eggs, dehydrated fruits and nuts (e.g., almond and prune), animal skin, tissues, etc.

Other than the fascinating self-assembled patterns and morphologies in biological systems, Mother Nature's strong power is witnessed even at the macroscopic geological systems. In many polar and high alpine environments, striking

Fig. 1.5 Forms of sorted patterned ground: (**a**) sorted circles and (**b**) sorted labyrinths, (**c**) sorted stripes, (**d**) sorted polygons (full scale bar: 1.0 m). (**e**) Rippling sand dunes under wind. Yin [1], reprinted with permission

circular, labyrinthine, polygonal, and striped patterns of stones and soil are spontaneously formed through self-organized processes (see Fig. 1.5a–d) [1], which constitute one of the most striking suites of geomorphic patterns. In sand dunes, the self-assembled meandering rippled pattern often observed in deserts is another beautiful scene in deserts [2], where the shape of ripples keeps across several thousand meters (see Fig. 1.5e).

1.2 Underlying Mechanical Mechanisms for Self-Assembled Pattern Formation

The self-assembled patterns and morphologies in Nature have fascinated and challenged both biologists and geomorphologists for decades [1, 3, 4]. For instance, the mechanisms underpinning the formation of fascinating shapes and patterns in plants have evoked scientists' interests for centuries [4]. While there is no doubt that biological and genetic factors significantly influence the morphogenesis, the active role of physics and mechanics should not be underemphasized, as stated by D'Arcy Thompson in his classical book on growth and forms [4],

> Cell and tissue, shell and bone, leaf and flower, are so many portions of matter, and it is in obedience to the laws of physics that their particles have been moved, moulded and conformed.

Thompson's argument on the inevitable interactions between physics and biology has been evidenced by recently increasing works, where the important role of mechanical force is found in the regulation of plant morphology [5, 6], cell growth and cell differentiation [7], and tissue morphogenesis [8], among others. Recently Hamant et al. [6] showed that mechanics plays a key active role in the development of plant organs, where the mechanical stress on plant tissue during growth controls the precise organization of a major structural element in the plant cell. By applying the growth hormone auxin to the edge of an eggplant leaf to cause a local expansion, Sharon et al. [9] showed that the growth stress makes the normally flat leaf buckle into a wavy one. Similar studies on the control of crinkly leaves were conducted by Nath et al. [10] even at the genetic level and they showed that flat leaves become crinkly in *Antirrhinum* after the *CIN* gene mutant, where leaves of *cin* cause excess growth in marginal regions and the resulting growth stress may lead to a curved surface. By investigating the undulation of a spherical Ag substrate/SiO_2 film system, Li et al. [11] successfully reproduced the Fibonacci number patterns in experiment that are similar to those observed in sunflowers' head, pine cones, and other plants, which suggested that mechanical force may be a driving force for some plant pattern formation.

For fruit morphology, from the mechanics point of view, the transition process from a normally smooth topology to a global undulating one (e.g., Fig. 1.3) may correspond to the occurrence of mechanical instability. Upon instability, the system will spontaneously choose the buckling mode with the lowest energy, often with an ordered self-assembled undulating pattern. Likewise, one may suspect that when the accumulated growth stress in the fruit arrives at the threshold, instability may occur and the spontaneous buckling pattern may underpin the distinctive morphologies observed in nature. This mechanical process may set a template for the subsequent and complex biological and biochemical processes (e.g., cell growth and differentiation) which may help to stabilize the distinctive patterns [12].

1.3 Bio-Inspired Self-Assembled Micro/Nanofabrication

Other than the spontaneously formed patterns in nature, in engineering, patterning has been developed as a technique for several decades to create functional and desired patterns and structures at the micro- and nanoscales. For example, in the field of microelectromechanical systems (MEMS), the traditional micro/nanofabrication techniques based on photolithography [13, 14] have been highly developed to fabricate micro- and nanopatterns and structures in the past 30 years [15]. The conventional photolithography approach uses UV light to transfer mode patterns to light-sensitive chemical photoresists on the underlying films or substrates, and after chemical etching patterned structures with high accuracy are produced. However, these methods are intrinsically expensive, time-consuming, effective to only a set of photoresists, and not directly applicable to nonplanar surfaces [16]. One of the central drawbacks is its limitation of fabricating small-aspect-ratio microstructures owing to the difficulty in deep etching methods [17]. In recent years, the *Lithographie, Galvanoformung, Abformung* (LIGA) technique has been developed to fabricate high-aspect-ratio microstructures to overcome the limitation [18]; however, it is low-efficient and highly expensive. The other main limitation of these traditional techniques is that the produced patterns and structures are inherently two dimensional (2D) as a result of the wafer-based fabrication methods. It is extremely challenging to create truly three-dimensional (3D) microstructures (e.g., on curved substrates) using lithography-based techniques. Thus the fabrication of 3D micro/nanopatterns and structures with high efficiency and low cost calls the need of revolutionary alternative approaches.

Among the conceptually new strategies offering possible routes to both small features and low costs, self-assembly involving thin films and multilayers has become the new cornerstone [16, 19]. The driving force for the thin film self-assembly can be the chemical interaction between film and substrate or physical forces such as magnetic, capillary, dispersion, and entropic interactions [16, 19]. Recently mechanical self-assembly of stiff film on planar substrate system driven by mismatched deformation has received wide applications in engineering [20], which offers a cost-effective and complementary solution to overcome the aforementioned challenges. The rich parameters that can be manipulated during the mechanical self-assembly process, including the material properties of film and substrate, substrate curvature, material and geometry gradient, mismatched stress, and time-dependent properties, etc., provide unlimited possibilities for the creation of varieties of functional 2D and 3D patterns and structures, with wide potential applications in biomedical engineering [21], optics [22], optoelectronics and display technologies [23, 24], among many others.

References

1. Yin J (2010) Mechanical self-assembly: science and applications. Ph.D. thesis, Columbia University
2. Partelia EJR, Duránb O, Tsoarc H, Schwämmled V, Herrmanna HJ (2009) Dune formation under bimodal winds. Proc Natl Acad Sci 106:22085
3. Darwin F (1897) The life and letters of Charles Darwin: including an autobiographical chapter. Appleton, New York
4. Thompson D (1992) On growth and form. Dover, New York
5. Fleming AJ, McQueen-Mason S, Mandel T, Kuhlemeier C (1997) Induction of leaf primordia by the cell wall protein expansin. Science 276:1415
6. Hamant O, Heisler MG, Jönsson H, Krupinski P (2008) Developmental patterning by mechanical signals in arabidopsis. Science 322:1650
7. Beloussov LV, Troshina TG, Mansurov AN (2008) Mechanical feedback in morphogenesis and cell differentiation. Biophysics 53:575
8. Ingber DE (2006) Mechanical control of tissue morphogenesis during embryological development. Int J Dev Biol 50:255
9. Sharon E, Marder M, Swinney HL (2004) Leaves, flowers and garbage bags: making waves. Am Sci 92:254
10. Nath U, Crawford BCW, Carpenter R, Coen E (2003) Genetic control of surface curvature. Science 299:1404
11. Li C, Zhang X, Cao Z (2005) Triangular and Fibonacci number patterns driven by stress on core/shell microstructures. Science 309:909
12. Yin J, Cao Z, Li C, Sheinman I, Chen X (2008) Stress-driven buckling patterns in spheroidal core/shell structures. Proc Natl Acad Sci USA 105:19132
13. Geppert L (1996) Semiconductor lithography for the next millennium. IEEE Spectrum 33:33
14. Jeong HJ, Markle DA, Owen G, Pease F, Grenville A, Bunau RV (1994) The future of optical lithography. Solid State Technol 37:39
15. Kovacs GTA, Maluf NI, Petersen KA (1998) Bulk micromachining of silicon. Proc IEEE 86:1536
16. Chen X, Yin J (2010) Buckling patterns of thin films on curved compliant substrates with applications to morphogenesis and three-dimensional micro-fabrication. Soft Matter 6:5667
17. Klaassen EH, Petersen K, Noworolski JM (1996) Silicon fusion bonding and deep reactive ion etching: a new technology for microstructures. Sensor Actuat A Phys 52:132
18. Malek CK, Saile V (2004) Applications of LIGA technology to precision manufacturing of high-aspect-ratio micro-components and -systems: a review. Microelectron J 35:131
19. Xia YN, Whitesides GM (1998) Soft lithography. Annu Rev Mater Sci 28:153
20. Bowden N, Brittain S, Evans AG, Hutchinson JW, Whitesides GM (1998) Spontaneous formation of ordered structures in thin films of metals supported on an elastomeric polymer. Nature 393:146
21. Jiang X, Bruzewicz DA, Wong AP, Piel M, Whitesides GM (2005) Directing cell migration with asymmetric micropatterns. Proc Natl Acad Sci 102:975
22. Chan EP, Crosby AJ (2006) Fabricating microlens arrays by surface wrinkling. Adv Mater 18:3238
23. Paul KE, Prentiss M, Whitesides GM (2003) Patterning spherical surfaces at the two-hundred-nanometer scale using soft lithography. Adv Funct Mater 13:259
24. Jin H, Abelson JR, Erhardt MK, Nuzzo RG (2004) Fabrication of a-Si:H photoconductive image sensor arrays on curved substrates using soft lithography. J Vac Sci Technol B 22:2548

Chapter 2
Mechanical Self-Assembly vs. Morphogenesis

Xi Chen and Jie Yin

Abstract Morphogenesis, as one of the three fundamental aspects of developmental biology, refers to the biological processes of developing certain shapes, which takes place across many length scales, including the morphologies of a cell, a tissue, an organ, and a system. From the intrinsic yet complicated biological and biochemical perspectives, several mechanisms for plant pattern formation have been suggested, such as positional information theory [1] and reaction–diffusion theory [2]. However, the active role of mechanical forces should not be underemphasized.

In the past few years, a great interest has been sparked in the development of biophysical and mechanical theories to explain the plant pattern formation [3, 4]. Among them, the connection of the morphogenetic processes of some plants with mechanical buckling theory receives a great attraction owing to some similarities. Patterns and shape formation are treated as the generation of specific undulating physical topography. From biophysical viewpoints, during the growth of plants, the morphology transition can be treated as spontaneously approaching the pattern/mode with minimal energy, which is similar to the mechanical instability/bifurcation approach. Among the several possible buckling/wrinkling modes (i.e., undulating patterns or structures), the system will spontaneously choose the pattern with the minimized energy.

X. Chen (✉)
Department of Earth and Environmental Engineering, Columbia Nanomechanics Research Center, Columbia University, New York, NY 10027, USA

SV Lab, International Center of Applied Mechanics, School of Aerospace, Xi'an Jiaotong University, Xi'an 710049, China

Department of Civil & Environmental Engineering, Hanyang University, Seoul 133-791, Korea
e-mail: xichen@columbia.edu

J. Yin
Department of Earth and Environmental Engineering, Columbia Nanomechanics Research Center, Columbia University, New York, NY 10027, USA

X. Chen (ed.), *Mechanical Self-Assembly: Science and Applications*,
DOI 10.1007/978-1-4614-4562-3_2, © Springer Science+Business Media New York 2013

2.1 Morphogenesis and Mechanical Buckling Model

A great interest has been sparked in the connection of the morphogenetic processes of some plants with mechanical buckling theory [3, 5–8]. Generally the related studies can be divided into two categories: one is the investigation on the mechanism of phyllotactic patterns in shoots and flowers; the other one is on the shape of leaves.

On phyllotactic patterns, Green pioneered the biophysical explanation for the patterns commonly observed in plant shoots and flowers [3, 4], and proposed the hypothesis that buckling of the compressed tunica is the governing mechanism for determining the local phyllotactic pattern. Dumais and Steele [6] showed that since the sunflower capitulum is under circumferential compression, buckling may be a plausible explanation for the primordium initiation in the captitulum. Shipman and Newell [5] demonstrated that the local phyllotaxis and the deformation configurations on plant surfaces may have resulted from the energy-minimizing buckling pattern of a compressed shell on an elastic foundation.

Several studies have been conducted on the shape of leaves from mechanical principles. Inspired by the similarity between wrinkled edges of torn plastic sheets and a wavy leaf, Marder et al. [9] suggested that some leaves may form wavy edges through spontaneous buckling and proposed a continuum theory on governing similar wavy edges observed in many leaves and flowers. To explain some wrinkled shapes in leaves, Dervaux and Amar [10] proposed an elasticity theory on the morphogenesis of growing tissues, where the growth stress is incorporated into the generalized Foppl–vonKarman (FvK) theory of thin plates. Koehl et al. [11] studied the ecomorphological differences on the blade shapes of kelp, where many species of macroalgae have wide, thin, and ruffled (undulate) blades in sheltered habitats, while their conspecifics at sites with more exposure to rapidly flowing water have narrow, thick, and flat blades. Their research revealed that the change in shape results from elastic buckling induced by mechanical stress. Recently, Liang and Mahadevan [12] studied the shape of long leaves and showed that the typical morphologies with saddle-like midsurface and rippled edges arise from the elastic relaxation via bending following differential growth in leaves. All these studies showed that the different shapes of leaves may result from the differential in-plane deformation within the leaves, which may lead to the occurrence of local or global wrinkling.

2.2 Mechanical Self-Assembly of Single-Layer Film and Multilayered Film–Substrate

In recent years, self-assembled fabrication of micro/nanopatterns and structures involving thin films and multilayers has become the new cornerstone, where there are numerous methods that enable self-assembly, such as surface-tension-based assembly [13, 14], electroactive polymer actuation [15], electric actuation [16], thermal and shape-memory alloy actuation [17], and stress-driven actuation

[18, 19]. In this dissertation, the studies related with mechanical self-assembly fabrication will be reviewed, especially by taking advantages of instability.

The mechanical self-assembly, i.e., the method of creating patterns or structures through utilizing failure of film, has become a focal point in the self-assembly fabrication. There are basically three kinds of failure in thin films: buckling [20], delamination [21], and cracking [22]. Among them, owing to the relative easiness of controlling and manipulation, the study on buckling of a single film sheet [23–26], bilayer thin films [27–30], and film–substrate systems [20, 31–33] has attracted wide interests and found extensive applications in engineering.

2.2.1 Mechanical Self-Assembly of Single- or Mutlilayer Film

Under mismatched deformation or external guidance, buckling may occur in a single-sheet film or multilayer films and leads to different morphologies. The controllable buckling process can be utilized and tailored to create desired patterns and structures. For a single sheet, there are two methods to trigger the self-assembly of a desired pattern or structure: one is the in-plane inhomogeneous and mismatched deformation at different locations; the other one is through the interaction between elasticity and capillarity, i.e., surface-tension-based assembly. For multilayer system, the driving force for the self-assembly mainly comes from the mismatched deformation between the different layers.

Through the programmed shrinkage at different locations of a thin flat gel sheet, Klein et al. [23] created a variety of both 3D large-scale buckling and multiscale wrinkling structures with nonzero Gauss curvatures as shown in Fig. 2.1. Although the resulting structures are not regular and ordered, it shows the promising of forming 3D structures from the buckling of a single 2D sheet.

Through the self-assembled wrapping of a liquid droplet by a planar sheet, Py et al. [24] developed a capillary origami of thin films into 3D structures as shown in Fig. 2.2. They showed that the final encapsulated 3D shapes can be controlled by tailoring the initial geometry of the flat membrane. Through the balance between interfacial energy and elastic bending energy, they revealed the critical length scale below which encapsulation cannot occur, which suggests a new way of mass production of 3D micro- or nanoscale structures. In a recent study it was demonstrated that the resulting 3D structures may offer a promising way to efficiently harvest solar energy in thin cells using concentrator microarrays [25]. A similar capillary origami study was conducted by Patra et al. [26] at the nanoscale using molecular dynamics simulations. They demonstrated that water nanodroplets may activate and guide the self-folding of planar grapheme nanostructures and lead to the self-assembly of nanoscale sandwiches, capsules, knots, and rings as shown in Fig. 2.3.

The mechanical bending and stretching ability of a 2D planar thin film, coupled with or without capillary driving force, may provide us an efficient self-assembly method to create folded 3D micro/nanostructures and devices, which have potential applications as building blocks of functional nanodevices, with unique mechanical, electrical, or optical properties [34].

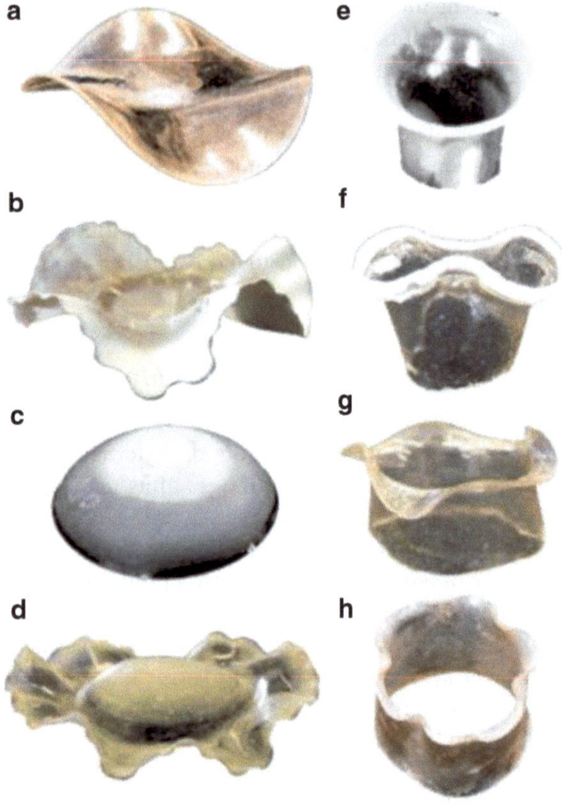

Fig. 2.1 Different structures of sheets with radially symmetric target metrics. Klein et al. [23], reproduced with permission

When combining another different thin film layer with the aforementioned single film sheet, different functional micro/nanostructures can be created through the mismatched deformation between the bilayer system, where such a mismatched deformation can be introduced by misfit lattice strain [29], different rates of thermal expansion [28], and swelling or contraction rate [16].

Schmidt and Eberl [29] pioneered the study on the self-assembly of thin solid films into nanotubes using misfit strain. They showed that when a bilayer of thin films with two different materials is deposited on substrates, after the bilayer is released by selective etching, the bilayer structure would buckle and bend upwards, and finally self-roll into a nanotube driven by the misfit lattice strain between the bilayers, as shown in Fig. 2.4a, b. Through the similar mismatched strain approach, Prinz et al. [35] further created more varieties of 3D micro/nanostructures such as tubes, coils, and helices with width ranging from a few micrometers down to a few nanometers (see Fig. 2.4c).

Despite the promise in creating nanotube or coils, Schmidt and Prinz et al.'s methods are limited to the large misfit lattice strain, which is only applicable to a few materials. By combining the advantages of tailored shape in Py et al.'s study and the self-rolling of tubes in Schmidt's study, recently Gracias and coworkers

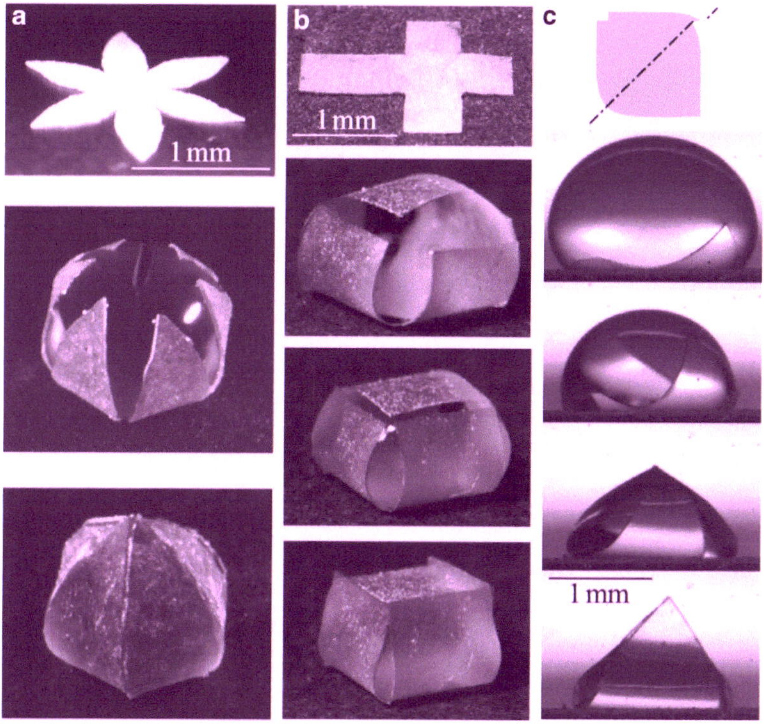

Fig. 2.2 Tuning of the initial flat shape to obtain (a) a spherical encapsulation, (b) a cubic encapsulation, or (c) a triangular mode-2 fold. Py et al. [24], reproduced with permission

[13, 27, 28] proposed and demonstrated a self-assembly method to design and fabricate complex patterned polyhedral micro-containers in the form of cubes, square pyramids, dodecahedra, and octahedral (see Fig. 2.5).

Their strategy uses the thin film sheets as a bilayer hinge consisting of a chromium (Cr)/copper (Cu) bilayer. The self-folding is caused by the residual stress developed during thermal evaporation of the metal thin films, which was due to the mismatch in the coefficient of thermal expansion (CTE) of the bilayer materials. Through the control of the thickness of each layer, the bilayer hinge could fold with certain desired angles and the resulting microstructures have great potential applications as vehicles for drug delivery [36] and 3D electromagnetic components.

2.2.2 Mechanical Self-Assembled Patterns in Film–Substrate System Through Wrinkling

Owing to the constraint of thin film thickness, the self-assembled 3D micro/nanostructures through single or bilayer thin films discussed above prefer the global

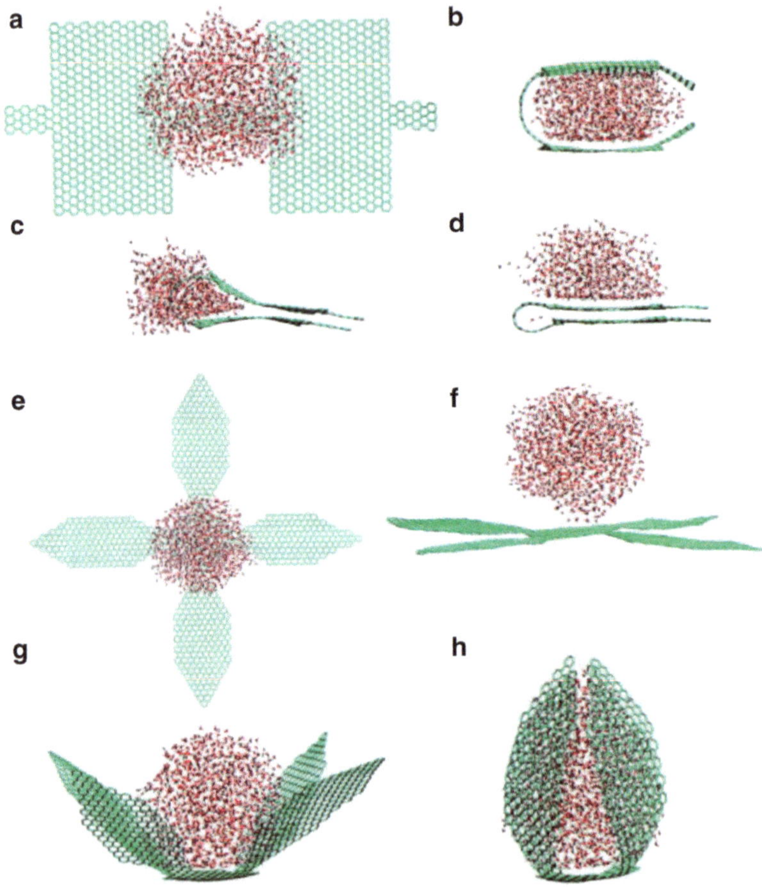

Fig. 2.3 (**a–d**) Water nanodroplet-activated and -guided folding of two graphene flakes connected by a narrow bridge. (**e–h**) Nanodroplet-assisted folding of a star-shaped graphene flake. Patra et al. [26], reproduced with permission

buckling to release the stored strain energy. It favors the global bending deformation with a large wavelength, which is analogous to the Euler buckling of columns. In the bucking of columns, the higher buckling modes with a shorter wavelength are not energetically unfavorable.

However, when the thickness of the underlying substrate is much larger than that of the thin film, since the substrate remains tightly bonded with the thin film during deformation, local wrinkling with a short wavelength will be preferred to release the compressive strain. The resulting wavelength is mainly determined by the competition between the bending energy of the thin film and the stretching energy of the substrates. While the bending energy prohibits the wrinkling with short wavelengths, the substrate favors wrinkles with shorter wavelengths. A trade-off between the bending energy in the film and stretching energy in the

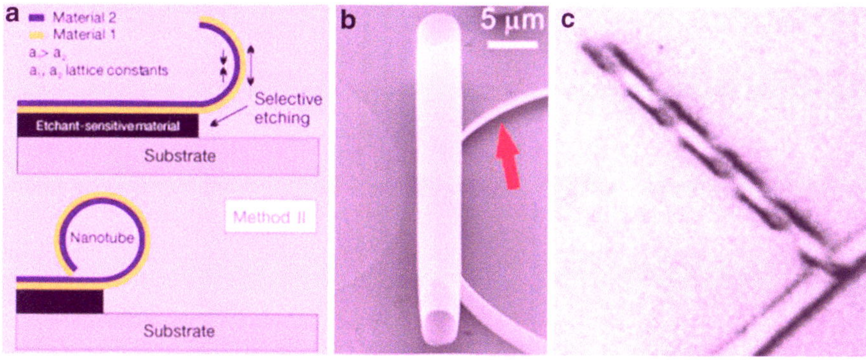

Fig. 2.4 Self-assembly of bilayer thin films (**a**) Schematic illustration of self-rolling of bilayer films under misfit lattice strain. (**b**) The resulting self-assembled nanotube. (**c**) Helix scrolled from a strip. Prinz et al. [35], reproduced with permission

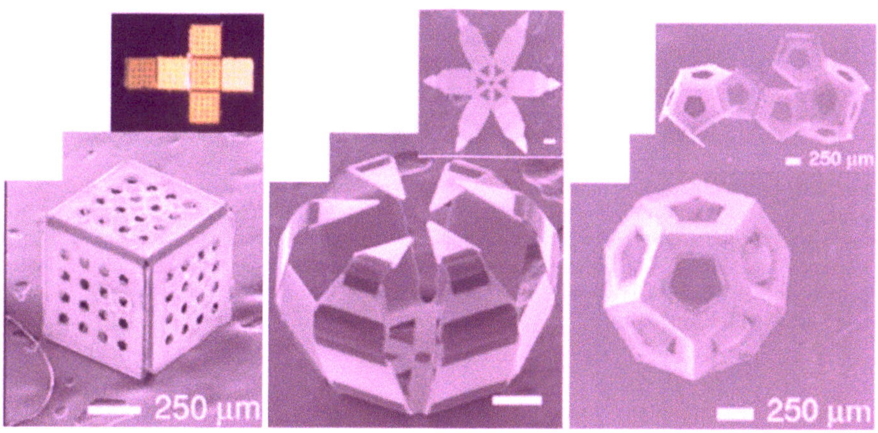

Fig. 2.5 Self-assembled patterned micro-containers through folding of bilayer hinges. On the right corner shows the original planar shape. Reprinted with permission from Bassik et al. [27]. Copyright 2009, American Institute of Physics

substrate determines the optimum wavelength of the wrinkles observed in film/substrate system.

Local wrinkling has been historically considered as a mechanism for structure failure, which should be avoided in the design of sandwich panels [37] and constructions widely used in aerospace and marine engineering [38] as well as the deposition of thin film in the semiconductor industry [39]. In 1969, Allen [37] first studied the problem on the wrinkling of sandwich panels in airplanes with a stiff face rested on a compliant substrate. The usefulness of spontaneous buckling of thin film/substrate systems was first demonstrated by Bowden et al. [20] in 1998, where in their pioneered work they utilized the wrinkling of metal films on soft polydimethylsiloxane (PDMS) substrates to generate ordered micropatterns and

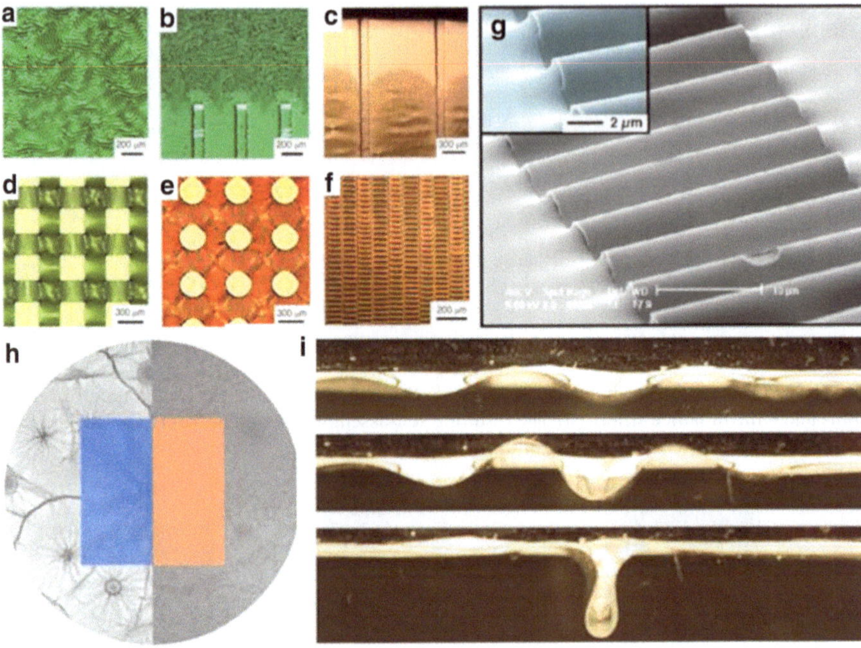

Fig. 2.6 Various wrinkling patterns observed in experiments. (**a–f**) Representative patterns in metal films on PDMS substrates through the control of edges and steps, where labyrinth patterns (**a**) far away any steps or edges transit to ordered ones through the rectangular, square, and circular steps or elevations. (**g**) One-dimensional ridged nanoribbons. (**h**) Concentric and radial wrinkled patterns through defects. (**i**) Wrinkle-to-fold transition of thin film on water. Moon et al. [51], reproduced with permission

structures. Spontaneous elastic buckling patterns were observed in the 50 nm-thick gold film as the system was cooled, owing to the mismatched thermal deformation and the films buckled into distinctive patterns, with typical wavelengths on the order of microns. These highly ordered patterns (e.g., Fig. 2.7a) can be precisely manipulated and have found vast applications in stretchable electronics [40–42], MEMS and NEMS [43], tunable optical gratings [44], thin film modulus measurement [45], force spectroscopy in cells [46], control of smart adhesion [47], adjustment of superhydrophobic properties of film [48], and pattern formation for micro/nanofabrication [20, 49] among others.

Following Bowden et al.'s pioneering experiment, extensive experimental, theoretical, and numerical studies were carried out to explore the buckling mechanisms and investigate the feasibility of quantitative control of the patterns for applications in micro- or nanostructures. Among experimental efforts, the substrate surface topology may be manipulated to change the local film stress, so as to generate a variety of ordered patterns (see Fig. 2.6a–f) [49].

Similarly, local physical properties of the thin film can be perturbed to result in various buckle patterns [50], and more refined nanoscale patterns may be achieved

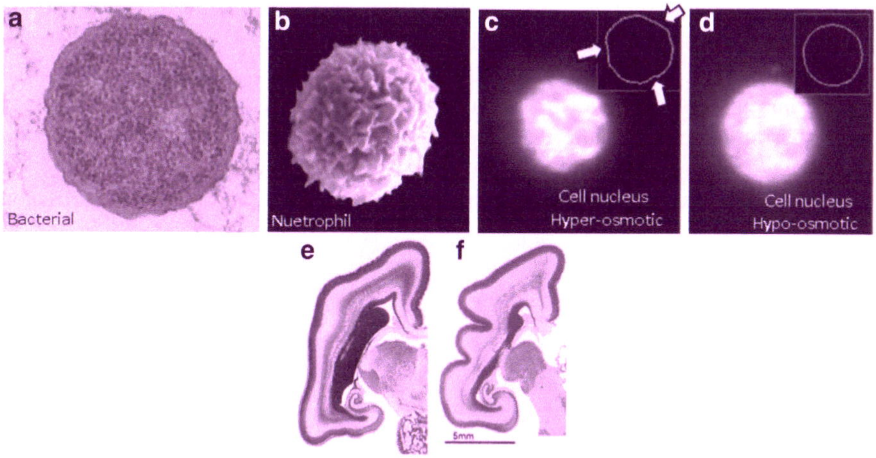

Fig. 2.7 The morphogenesis of some cells and tissues may be related to the wrinkling instability of nearly spherical shell/core systems. *For cells*: (**a**) Wrinkled bacterial cell owing to the relative shrinkage of the cytoplasm under hyper-osmotic pressure. (**b**) Wrinkled human neutrophil cell due to the relative expansion of the cell membrane surface area during cell growth or phagocytosis. (**c**) Wrinkled cell nucleus due to hyper-osmotic shrinkage, and (**d**) the wrinkles may disappear with the swelling of nucleoplasm under hypo-osmotic pressure. *The folding pattern of brain cortex*: the cross-sectional view of brain cortex shows that (**e**) during the early stage the surface is relatively smooth, and (**f**) during the later stage the wrinkled morphology is observed. Chen and Yin [76], reproduced with permission

by modifying the surfaces using focused ion beam [51]. External constraints may be applied, where a pre-patterned mold was held against the film as the buckles were formed, and the resulting pattern was quite stable after the removal of the mold [52]. The substrate may also be pre-strained [53], where silicon nano-ribbons bonded to a pre-stretched flat PDMS could generate wavy layouts upon releasing of the substrate strain (see Fig. 2.7g). Other than the discovered labyrinth, one-dimensional wavy and herringbone patterns in thin films, recently Chung et al. [32] demonstrated new types of wrinkles in a dendritic-like spoke pattern or in a target pattern consisting of concentric rings near defects in films (see Fig. 2.7h). Besides solid substrates, a recent work [54] reported a wrinkle-to-fold transition of buckled thin polyester film on water or soft gel substrates (see Fig. 2.7i).

In order to explain the formation mechanisms of various intriguing buckling patterns from theoretical aspects, Cerda and Mahadevan [55] proposed a generalized scaling law for the buckling wavelength and amplitude; Chen and Hutchinson [31, 56] showed that upon equi-biaxial compression of a film bonded to a planar semi-infinite compliant substrate, the herringbone pattern possesses less strain energy than its competitors and thus is more favorable. A more comprehensive discussion of the herringbone mode was recently given by Audoly and Boudaoud [57]. Huang et al. [58] analytically investigated the substrate thickness on the wrinkling wavelength and proved the feasibility of neglecting interface shear stress in theory.

By studying the kinetic buckling of elastic films on viscoelastic substrates, Huang [59] showed that both energetics and kinetics play important roles in determining the critical condition, the growth rate, and the wavelength. Through the control of anisotropic strains in films, the evolution and transition of stripes, herringbone, and labyrinth buckling patterns were simulated [60]. In a review, Genzer and Groenewold gave extensive examples of patterns achievable via film wrinkling and bridged that with skin wrinkles and possible ways of material characterization and fabrication.

2.2.3 *Mechanical Self-Assembly (MSA) of Thin Film on Curved Substrate*

The buckling characteristics of closed thin film (shell) on *curved* substrate (core) have important implications in the morphogenesis of quite a few fruits, vegetables, fingertips, animal skins, tissues, and cells [61–64] as discussed earlier. The intriguing wrinkling-like ordered patterns observed in these systems may be related to mechanics-driven buckling process owing to the mismatched deformation between the shell/film (e.g., skin of fruit or membrane of cell) and the underlying curved core/substrate (e.g., flesh of fruit or cytoplasm of cell), during which the curvature of the substrate may play a dominant role in shaping the distinctive overall appearance of quite a few natural and biological systems.

On morphogenesis, wrinkled cells are often observed in bacterial cells (Fig. 2.7a where the average wrinkle wavelength is about 100 nm) [65] and non-tissue cells such as human neutrophil cells (Fig. 2.7b) [66], macrophages, lymphocytes [67], and mast cells [68]. The wrinkled morphology may increase the surface area of the cell by more than 100 % [69], which may accommodate potential membrane expansion and spreading during extravasation and osmotic swelling. Other than the cell membrane surface wrinkles, recently similar wrinkled morphology was observed inside the cell, e.g., on the cell nucleus due to hyper-osmotic shrinkage, Fig. 2.7c [70]. In their work, Finan and Guilak [70] suggested that the nucleus wrinkles can be explained by the mechanical buckling of shell/core structure, where the contraction of the soft core (nucleoplasm) renders the stiffer shell (nuclear lamina) in compression to initiate the buckles. Under hypo-osmotic pressure, the swelling of nucleoplasm will make the lamina in tension and stretch the lamina into a smooth shape as shown in Fig. 2.7d [70].

Besides the cellular scale, the wrinkled morphology is also frequently observed at larger tissue or organ scales. The wrinkled brain cortex shown in Fig. 1.1c, d is a good example of wrinkling on curved substrates. The cross-section of the hemispherical cortex reveals the detailed information on the formation of the gyri (ridge) and sulci (groove) during development (see Fig. 2.7e, f) [71]. A number of hypotheses and models were proposed to explain how and why the cortex folds in a characteristic pattern from the biological, biochemical, and mechanical

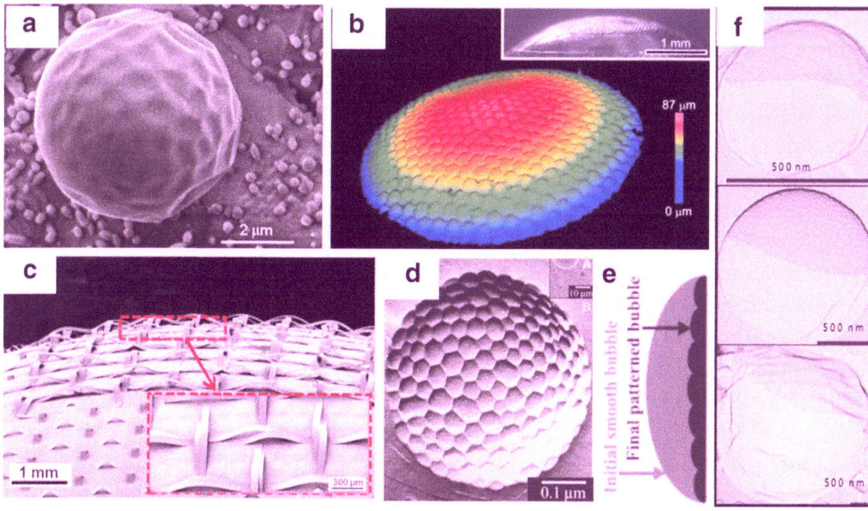

Fig. 2.8 Experiments of self-assembly on spherical shell/core systems. *Solid inorganic systems*: (**a**) Experiment of reticular pattern formed on a spherical system (SiO_2 film/Ag substrate). (**b**) Microlens arrays self-assembled on a hemispherical soft substrate using constrained local buckles. (**c**) Interconnected silicon ribbonlike photodetectors on a hemispherical elastomer substrate. *Instability patterns of microbubbles*: (**d**) Nanoscale hexagonal pattern self-assembled on a stable microbubble, which is in part due to (**e**) differential shrinkage-induced buckling of the bubble surface, and (**f**) the pattern can be strongly influenced by the bubble curvature. Chen and Yin [76] reproduced with permission

viewpoints [72–76]. Among them, Richman et al. [72] first proposed a mechanical buckling model where the cerebral cortex was modeled as a bilayer shell rested on a soft spherical core. The excessive growth of the shell relative to that of the core leads to the development of compressive stress in the shell, and the subsequent buckling may lead to the cortical folding.

Moreover, thanks to the rapid development of soft lithography which enables micro- and nanofabrication of multilevel thin-film devices with nonplanar geometries [77, 78], controlled 3D patterning on curved and/or closed substrates could significantly expand applications in biomedical engineering [79], optics [50], optoelectronics, and display technologies [80–82]. By investigating the undulation of a spherical Ag substrate/SiO_2 film system, Cao et al. [83] demonstrated an experiment of spontaneous buckling pattern formation on spherical substrates. Figure 2.8a shows an example with substrate radius $R = 3$ μm and at this relatively small R/t with $t = 150$ nm being the film thickness, reticular pattern was produced via spontaneous buckling. This serves an example of buckling self-assembly fabrication of true 3D structures at micro or submicron scales. To create 3D micro/nanopatterns and microstructures on curved substrates, Chan and Crosby [50] confined surface wrinkles in small local pre-patterned regions, and when such a technique was applied to a hemispherical surface, microlens arrays were self-assembled (Fig. 2.8b).

Recently, Shin et al. [84] assembled interconnected silicon ribbonlike photodetectors on a hemispherical elastomer substrate, and the ribbons were in buckled profiles owing to pre-stretch (Fig. 2.8c). It should be noted that in these experiments, the buckle features created were much smaller than the substrate radius of curvature; in other words, the versatile effect and potential of substrate curvature were not fully utilized to regulate the self-assembled buckles. Besides the aforementioned solid systems, buckling self-assembly was also demonstrated on spherical shell with hollow core microstructures (fluidic spherical shell/core microstructures), i.e., microbubbles. In a recent experiment on the nanopatterning of stable microbubbles, Dressaire et al. [85] created a nanoscale hexagonal interface pattern as shown in Fig. 2.8d through the shrinkage of the bubbles. By covering a surfactant layer on the surface of microbubbles, due to the differential shrinkage the initial smooth bubble buckled into a nano-hexagonal patterned one (Fig. 2.8e). Figure 2.8f further demonstrates the important effect of curvature on the surface wrinkling pattern, as the bubble radius was varied from 500 nm to 3 mm. Note that besides mechanical buckling, phase separation and other surface mechanisms may also underpin the pattern domains in microbubbles, and various pattern formations were reported including polygons, dendrites, beans, networks, etc. [86]. A recent review by Borden [86] has nicely summarized the nanopatterning on stable microbubbles, which have implications for biomedical applications.

References

1. Held L (1992) Models for embryonic periodicity. Karger Press, New York
2. Harrison LG (1993) Kinetic theory of living pattern. Cambridge University Press, Cambridge, UK
3. Green PB (1999) Expression of pattern in plants: combining molecular and calculus-based biophysical paradigms. Am J Bot 86:1059
4. Green PB, Steele CS, Rennich SC (1996) Phyllotactic patterns: a biophysical mechanism for their origin. Ann Bot 77:512
5. Shipman PD, Newell AC (2004) Phyllotactic patterns on plants. Phys Rev Lett 92:168102
6. Dumais J, Steele CR (2000) New evidence for the role of mechanical forces in the shoot apical meristem. J Plant Growth Regul 19:7
7. Sharon E, Marder M, Swinney HL (2004) Leaves, flowers and garbage bags: making waves. Am Sci 92:254
8. Steele CR (2000) Shell stability related to pattern formation in plants. J Appl Mech 67:237
9. Marder M, Sharon E, Smith S, Roman B (2003) Theory of edges of leaves. Europhys Lett 62:498
10. Dervaux J, Amar MB (2008) Morphogenesis of growing soft tissues. Phys Rev Lett 101:068101
11. Koehl MAR, Silk WK, Liang H, Mahadevan L (2008) How kelp produce blade shapes suited to different flow regimes: a new wrinkle. Integr Comp Biol 48:834
12. Liang H, Mahadevan L (2009) The shape of a long leaf. Proc Natl Acad Sci 106:22049
13. Leong TG, Lester PA, Koh TL, Call EK, Gracias DH (2007) Surface tension-driven self-folding polyhedra. Langmuir 23:8747
14. Syms RRA, Yeatman EM, Bright VM, Whitesides GM (2003) Surface tension-powered self-assembly of microstructures—the state of the art. J Microelectromech Syst 12:387

15. Jager EWH, Smela E, Inganas O (2000) Microfabricating conjugated polymer actuators. Science 290:1540
16. Suzuki K, Yamada H, Miura H, Takanobu H (2007) Self-assembly of three dimensional micro mechanisms using thermal shrinkage of polyimide. Microsyst Technol 13:1047
17. Luo JK, He JH, Fu YQ, Flewitt AJ, Spearing SM, Fleck NA, Milne WI (2005) Fabrication and characterization of diamond-like carbon/Ni bimorph normally closed microcages. J Micromech Microeng 15:1406
18. Arora WJ, Nichol AJ, Smith HI, Barbastathis G (2006) Membrane folding to achieve three-dimensional nanostructures: nanopatterned silicon nitride folded with stressed chromium hinges. Appl Phys Lett 88:053108
19. Moiseeva E, Senousy YM, McNamara S, Harnett CK (2007) Single-mask microfabrication of three-dimensional objects from strained bimorphs. J Micromech Microeng 17:N63
20. Bowden N, Brittain S, Evans AG, Hutchinson JW, Whitesides GM (1998) Spontaneous formation of ordered structures in thin films of metals supported on an elastomeric polymer. Nature 393:146
21. Yu HH, He MY, Hutchinson JW (2001) Edge effects in thin film delamination. Acta Mater 49:93
22. Ye T, Suo Z, Evans AG (1992) Thin film cracking and the roles of substrate and interface. Int J Solids Struct 29:2639
23. Klein Y, Efrati E, Sharon E (2007) Shaping of elastic sheets by prescription of non-Euclidean metrics. Science 315:1116
24. Py C, Reverdy P, Doppler L, Bic J, Roman B (2007) Capillary origami: spontaneous wrapping of a droplet with an elastic sheet. Phys Rev Lett 98:156103
25. Guo X, Li H, Ahn BY, Duoss EB, Hsia KJ, Lewis JA, Nuzzo RG (2009) Two- and three-dimensional folding of thin film single-crystalline silicon for photovoltaic power applications. Proc Natl Acad Sci 106:20149
26. Patra N, Wang B, Kral P (2009) Nanodroplet activated and guided folding of graphene nanostructures. Nano Lett 9:3766
27. Bassik N, Stern GM, Gracias DH (2009) Microassembly based on hands free origami with bidirectional curvature. Appl Phys Lett 95:091901
28. Leong TG, Benson BR, Call EK, Gracias DH (2008) Thin film stress driven self-folding of microstructured containers. Small 4:1605
29. Schmidt OG, Eberl K (2001) Thin solid films roll up into nanotubes. Nature 410:168
30. Huang M, Boone C, Roberts M, Savage DE, Lagally MG, Shaji N (2005) Nanomechanical architecture of strained bilayer thin films: from design principles to experimental fabrication. Adv Mater 17:2860
31. Chen X, Hutchinson JW (2004) Herringbone buckling patterns of compressed thin films on compliant substrates. J Appl Mech 71:597
32. Chung JY, Nolte AJ, Stafford CM (2009) Diffusion-controlled, self-organized growth of symmetric wrinkling patterns. Adv Mater 21:1
33. Genzer G, Groenewold J (2006) Soft matter with hard skin: from skin wrinkles to templating and material characterization. Soft Matter 2:310
34. Baughman RH, Cui CX, Zakhidov AA (1999) Carbon nanotube actuators. Science 284:1340
35. Prinz VY, Seleznev VA, Gutakovsky AK, Chehovskiy AV (2000) Free-standing and overgrown InGaAs/GaAs nanotubes, nanohelices and their arrays. Physica E 6:828
36. Freitas RA (1999) Nanomedicine. Landes Bioscience, Georgetown, TX
37. Allen HG (1969) Analysis and design of structural sandwich panels. Pergamon, New York
38. Imhof B (1997) Modern sandwich core materials offer new attractive design possibilities. Aircraft Eng Aerosp Technol 69:332
39. Freund LB, Suresh S (2003) Thin film materials: stress, defect formation and surface evolution. Cambridge University Press, Cambridge (UK)
40. Sun Y, Choi WM, Jiang H, Huang Y, Rogers JA (2006) Controlled buckling of semiconductor nanoribbons for stretchable electronics. Nat Nanotechnol 1:201

41. Khang DY, Jiang HQ, Huang Y, Rogers JA (2006) A stretchable form of single-crystal silicon for high-performance electronics on rubber substrates. Science 311:208
42. Lacour SP, Wagner S, Huang ZY, Suo Z (2003) Stretchable gold conductors on elastomeric substrates. Appl Phys Lett 82:2404
43. Fu YQ, Sanjabi S, Barber ZH, Clyne TW (2006) Evolution of surface morphology in TiNiCu shape memory thin films. Appl Phys Lett 89:171922
44. Efimenko K, Rackaitis M, Manias E, Vaziri A, Mahadevan L, Genze J (2005) Nested self-similar wrinkling patterns in skins. Nat Mater 4:293
45. Stafford CM, Harrison C, Beers KL, Karim A, Amis EJ, Vanlandingham MR, Kim HC, Volksen W, Miller RD, Simonyi EE (2004) A buckling-based metrology for measuring the elastic moduli of polymeric thin films. Nat Mater 3:545
46. Harris AK, Wild P, Stopak D (1980) Silicone rubber substrata: a new wrinkle in the study of cell locomotion. Science 208:177
47. Chan EP, Smith E, Hayward R, Crosby AJ (2008) Surface wrinkles for smart adhesion. Adv Mater 20:711
48. Lin P, Yang S (2009) Mechanically switchable wetting on wrinkled elastomers with dual-scale roughness. Soft Matter 5:1011
49. Bowden N, Huck WTS, Paul KE, Whitesides GM (1999) The controlled formation of ordered, sinusoidal structures by plasma oxidation of an elastomeric polymer. Appl Phys Lett 75:2557
50. Chan EP, Crosby AJ (2006) Fabricating microlens arrays by surface wrinkling. Adv Mater 18:3238
51. Moon MW, Lee SH, Sun JY, Oh KH, Vaziri A, Hutchinson JW (2007) Controlled formation of nanoscale wrinkling patterns on polymers using focused ion beam. Scr Mater 57:747
52. Yoo PJ, Suh KY, Lee HH (2002) Physical self-assembly of microstructures by anisotropic buckling. Adv Mater 14:1383
53. Jiang HQ, Khang D, Song J, Sun Y, Huang Y, Rogers JA (2007) Finite deformation mechanics in buckled thin films on compliant supports. Proc Natl Acad Sci 104:15607
54. Pocivavsek L, Dellsy R, Kern A, Johnson S, Lin B, Lee KYC, Cerda E (2008) Stress and fold localization in thin elastic membranes. Science 320:912
55. Cerda E, Mahadevan L (2003) Geometry and physics of wrinkling. Phys Rev Lett 90:074302
56. Chen X, Hutchinson JW (2004) A family of herringbone patterns in thin films. Scr Mater 50:797
57. Audoly B, Boudaoud A (2008) Buckling of a stiff film bound to a compliant substrate-Part II: a global scenario for the formation of herringbone pattern. J Mech Phys Solids 56:2422
58. Huang ZY, Hong W, Suo Z (2005) Nonlinear analyses of wrinkles in films on soft elastic substrates. J Mech Phys Solids 53:2101
59. Huang R (2005) Kinetic wrinkling of an elastic film on a viscoelastic substrate. J Mech Phys Solids 53:63
60. Huang Z, Hong W, Suo Z (2004) Evolution of wrinkles in hard films on soft substrates. Phys Rev E Stat Nonlin Soft Matter Phys 70:030601
61. Kuhl E, Maas R, Himpel G, Menzel A (2007) Computational modeling of arterial wall growth. Biomech Model Mechanobiol 6:321
62. Hallett MB, von Ruhland CJ, Dewitt S (2008) Chemotaxis and the cell surface-area problem. Nat Rev Mol Cell Biol 9:662
63. Wang B, Liu P, Jiang W, Pan H, Xu X, Tang R (2008) Yeast cells with an artificial mineral shell: protection and modification of living cells by biomimetic mineralization. Angew Chem Int Ed 47:3560
64. Rahmy TR, Ayoub MA (2002) In vitro action of cobra venom on goat spermatozoa ultrastructure by transmission and scanning electron microscopy. J Venom Anim Toxins 8:127
65. Schwarz H, Koch AL (1995) Phase and electron microscopic observations of osmotically induced wrinkling and the role of endocytotic vesicles in the plasmolysis of the gram-negative cell wall. Microbiology 141:3161
66. Bessis M (1973) Living blood cells and their ultrastructure. Springer, Berlin

67. Tohya K, Kimura M (1998) Ultrastructural evidence of distinctive behavior of L-selectin and LFA-1 alpha (L) beta (2) integrin on lymphocytes adhering to the endothelial surface of high endothelial venules in peripheral lymph nodes. Histochem Cell Biol 110:407
68. Burwen SJ, Satir BH (1977) Plasma membrane folds on the mast cell surface and their relationship to sectretory activity. J Cell Biol 74:690
69. Dewitt S, Hallett M (2007) Leukocyte membrane "expansion": a central mechanism for leukocyte extravasation. J Leukoc Biol 81:1160
70. Finan JD, Guilak F (2010) The effects of osmotic stress on the structure and function of the cell nucleus. J Cell Biochem 109:460
71. Bayer S, Altman J (2005) The human brain during the second trimester. Taylor & Francis, Boca Raton
72. Richman DP, Stewart RM, Hutchinson JW, Caviness VSJ (1975) Mechanical model of brain convolutional development. Science 189:18
73. Essen DCV (1997) A tension-based theory of morphogenesis and compact wiring in the central nervous system. Nature 385:313
74. Armstrong CL, Hawkes R (2000) Pattern formation in the cerebellar cortex. Biochem Cell Biol 78:551
75. Geng G, Johston LA, Yan E, Britto JM, Smith DW, Walker DW, Egan GF (2009) Biomechanisms for modelling cerebral cortical folding. Med Image Anal 13:920
76. Chen X, Yin J (2010) Buckling patterns of thin films on curved compliant substrates with applications to morphogenesis and three-dimensional micro-fabrication. Soft Matter 6:5667
77. Erhardt MK, Jin H, Abelson JR, Nuzzo RG (2000) Low-temperature fabrication of Si thin-film transistor microstructures by soft lithographic patterning on curved and planar substrates. Chem Matter 12:3306
78. Xia YN, Whitesides GM (1998) Soft lithography. Annu Rev Mater Sci 28:153
79. Jiang X, Bruzewicz DA, Wong AP, Piel M, Whitesides GM (2005) Directing cell migration with asymmetric micropatterns. Proc Natl Acad Sci 102:975
80. Ruchhoeft P, Colburn M, Choi B, Nounu H, Johnson S, Bailey T, Damle S, Stewart M, Ekerdt J, Sreenivasan SV, Wolfe JC, Willson CG (1999) Patterning curved surfaces: template generation by ion beam proximity lithography and relief transfer by step and flash imprint lithography. J Vac Sci Technol B 17:2965
81. Jin H, Abelson JR, Erhardt MK, Nuzzo RG (2004) Fabrication of a-Si:H photoconductive image sensor arrays on curved substrates using soft lithography. J Vac Sci Technol B 22:2548
82. Paul KE, Prentiss M, Whitesides GM (2003) Patterning spherical surfaces at the two-hundred-nanometer scale using soft lithography. Adv Funct Mater 13:259
83. Cao G, Chen X, Li C, Ji A, Cao Z (2008) Self-assembled triangular and labyrinth buckling patterns of thin films on spherical substrate. Phys Rev Lett 100:036102
84. Shin G, Jung I, Malyarchuk V, Song J, Wang S, Ko HC, Huang Y, Ha JS, Rogers JA (2010) Micromechanics and advanced designs for curved photodetector arrays in hemispherical electronic-eye cameras. Small 6:851
85. Dressaire E, Bee R, Bell DC, Lips A, Stone HA (2008) Interfacial polygonal nanopatterning of stable microbubbles. Science 320:1998
86. Borden M (2009) Nanostructural features on stable microbubbles. Soft Matter 5:716

Chapter 3
Shaping by Active Deformation of Soft Elastic Sheets

Eran Sharon

3.1 Introduction

One shaping method of soft elastic bodies is growth. This method, which is very common in natural systems and much less in man-made structures, is inherently different from shaping by external loading. In the second case the body responses *passively* to the constraints imposed on its boundaries. It has a reference stress-free configuration and the task is to compute and minimize the elastic energy of configurations that fulfill the boundary conditions. In the case of self-shaping by growth, the body is free in space, but contains a field of *active* growth, or deformation, which determines the shape of the body.

If we want to turn shaping via growth into a controllable design technique, there are several important questions that must be addressed: How can we describe the physics of shaping via growth? What is the correct and convenient way of integrating the information about active growth into the theory of elasticity? In many cases, or at least in the interesting ones, growing bodies do not have a stress-free configuration. What is the reference quantity with respect to which the elastic energy can be computed and minimized? Which materials are suitable for the construction of self-shaping bodies? How can we build bodies with predetermined growth distribution and how can we trigger this growth? From a different perspective we can ask if and how the physics of shaping via growth is used in natural growing bodies. Can we point out examples where the principles of the mechanics of growth play a key role in symmetry breaking during morphogenesis? And finally, can we use these principles to affect and correct morphogenesis, where it is needed?

In this chapter we try to provide some suggestions and guidelines for suitable theoretical framework and experimental techniques. We provide some specific examples that illustrate central principles that govern this shaping method in

E. Sharon (✉)
The Racah Institute of Physics, The Hebrew University of Jerusalem, Jerusalem, Israel
e-mail: erans@vms.huji.ac.il

X. Chen (ed.), *Mechanical Self-Assembly: Science and Applications*,
DOI 10.1007/978-1-4614-4562-3_3,

synthetic and natural systems. It is important to say that this entire field is in its early stages. New concepts, materials, and techniques are likely to be developed in the near future and to extend the applicability and implementation of this new branch of mechanics.

3.2 Theoretical Framework

3.2.1 Growth of Elastic Bodies

First we need to clarify what do we mean by a "growing elastic body." We should think of an elastic material, in which "rest" (in the sense of energy minimum) distances between neighboring points are given at any time, as if they are connected by springs. When we say that such a material grows we mean that these rest distances are slowly changed in time—the rest lengths of the springs are redetermined, but they continue to response elastically to deformations around the new rest length. The laws governing the growth process are outside of elasticity theory and we treat them as given (unlike the case of elasto-plasticity theories). Though extremely interesting and important, we do not attempt to say anything about these laws. We just assume that we know their outcome, i.e., the rest length of all the "springs." In addition, we assume that the growth is much slower than the elastic response and we study only equilibrium configurations.

As a result of growth, the body, which is free of external constraints, is deformed and we would like to use the theory of elasticity in order to calculate its equilibrium configuration. As in conventional elasticity, we would like to write an energy functional and to find the configuration that minimizes it.

3.2.2 Residually Stressed Incompatible Bodies

In the theory of elasticity, the elastic energy functional is a volume integral of an energy density, which is expressed by the local strain tensor and material elastic properties [1, 2]. In most formulations of elasticity the strain tensor of a given configuration is expressed by gradients of the map between a stress-free configuration of a body and its current configuration. When we wish to calculate the strain within a buckled compressed ruler we compute the displacement field that takes every point in the ruler, from its stress-free, flat, configuration to the buckled configuration. Derivatives of the displacement field tell us by how much distance between neighboring points has changed, i.e. by how much the spring connecting them is stretched.

Perhaps surprisingly, this basic procedure for definition of strain does not work in the general case of growing elastic bodies.

Consider an elastic ball of radius R and let the material within $r<R/2$ grow isotropically by 10 %, while for $R/2<r<R$, the material does not grow at all. One immediately sees that there is no way by which the inner ball and the outer shell can fit together while keeping their new equilibrium distances everywhere. Some parts of the grown body must be stretched and some parts will be compressed even if the ball is free in space. The grown ball is residually stressed—it does not have any stress-free configuration! The lack of stress-free configuration is not a consequence of the sharp difference between the two parts of the ball in the example. Even a continuously varied growth field could lead to a similar situation [3, 4]. Having no stress-free configuration, how can we define an elastic strain for such a body?

3.2.3 Metric Description

Considering the example above, one immediately feels that the difficulties in defining an elastic strain are formal or technical, but not "physical." If, as assumed, we know the rest length of each spring, we can compare it to its length in every given configuration. So where is the problem? We note that when defining strain by the displacement field from a stress-free configuration, we "*simultaneously*" compare the lengths of all springs in a configuration to their rest lengths in the stress-free *configuration*. In fact, we do not have to do so. We do not care if there is a configuration in which all springs are simultaneously at their rest length. We just need to properly express these lengths and compare them to those in a given configuration.

A natural geometric quantity which describes local lengths in a body is the metric tensor [5]. It provides a measure of distances in different directions at every point in the body and can be expressed "extrinsically" as a property of a configuration, or intrinsically without the need of a realization as a "body." When considering an elastic body, the "lengths of springs" in a given configuration are expressed by the metric tensor field $g(r)$ of the configuration. Equivalently, the rest lengths of springs in a body can be expressed by *a reference metric* $\overline{g}(r)$. The reference metric, $\overline{g}(r)$, contains all the information which is determined by the growth, i.e., the *rest lengths* in each direction at each point in the body, while the actual metric, $g(r)$, expresses the *actual lengths* at the same points. Realizing this the elastic strain tensor can be written as

$$\varepsilon(r) = \frac{1}{2}(g(r) - \overline{g}(r)). \tag{3.1}$$

This definition of strain holds even if the reference metric $\overline{g}$ cannot be realized as a configuration in space. In addition, when there is a stress-free configuration, $\overline{g}$ is its metric and the strain in (3.1) coincides with the conventional definition of the (nonlinear) strain (see [3]).

Turning to the elastic energy, we note that since the elastic energy density is a local scalar quantity, we can compute it "separately" for every volume element (or "spring") using the strain in (3.1) and the material elastic coefficients (assuming linear stress–strain relation), and then sum it up:

$$E \sim \int A(r)(g - \overline{g})^2 \mathrm{d}v. \tag{3.2}$$

$A(r)$ is a tensor containing the local elastic properties. For isotropic homogeneous material it has only two independent parameters—the Lame coefficients (see [2]). Using the above expression it is important to note that the volume element $\mathrm{d}v$ is calculated correctly and consistently (see [3]).

To summarize: Growth of an elastic body determines a reference metric $\overline{g}$ on the body. Every configuration of the body has an actual metric g. The local elastic strain is the difference between g and $\overline{g}$ (3.1) and the energy is an integral over the energy density, which is quadratic in the elastic strain (3.2).

3.3 Thin Sheets

In the rest of this chapter we limit our discussion to thin growing elastic sheets—structures with one dimension—the thickness, t (in the z direction), much smaller than the other two (x- and y-directions).

First, we will look at such ordinary (not growing) bodies. For such bodies we can define their mid surface, $z = 0$, and their upper and lower "faces" at $\pm t/2$. The thinness of these bodies allowed the derivation of *two-dimensional (2D) effective theories*—the theories of plates and shells [6–8]. In these theories a configuration of a sheet is determined by the shape of its 2D mid surface. To complete the state of the 3D body, the variation of strain in the z-direction is assumed to be linear with the distance from the mid surface, and the z component of the strain is assumed to vanish [9]. Under these conditions the energy of a thin sheet takes the general form of a sum of stretching and bending energies:

$$E = E_s + E_b.$$

The stretching energy is quadratic in the in-plane strain on the mid surface and linear in the thickness. The bending energy stems from the nonuniform "horizontal" strain in "layers" below and above the mid plane. This strain is assumed (or approximated) linear in z. Integrating the elastic energy density (which is quadratic in z) across the thickness leads to the bending energy, which is cubic in t and quadratic in the deviation of the surface curvature k from its natural or "spontaneous" curvature k_0. We distinguish between two main types of sheets: *Shells*, with nonzero spontaneous curvature, $k_0 \neq 0$, and *plates* for which $k_0 = 0$ (Fig. 3.1).

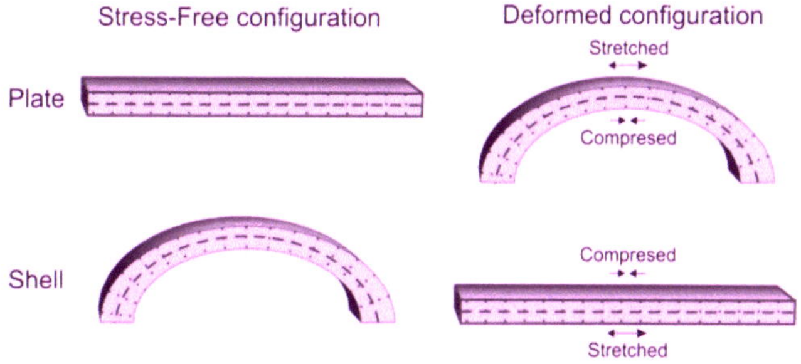

Fig. 3.1 Schematics of a plate and a shell with and without a pure bending deformation, with mid surface (*dashed line*) and its normals (*dotted lines*) indicated. A stress-free plate (*top left*) is flat—has no spontaneous curvature. When bent (*top right*) its upper "layers" are stretched and lower layers are compressed. A stress-free shell (*bottom left*) is not flat. It has a spontaneous curvature (in this illustration it is uniaxial, but in general it is tensorial). When bent (for example flattened, as in the bottom right panel) upper/lower layers are stretched/compressed, depending on the bending direction. Most effective 2D models assume that normals to the mid surface remain normals in the deformed configuration as illustrated

3.3.1 Metric Description of Thin Sheets

Converting the 3D metric elasticity into effective 2D plate/shell theories follows basically the same steps as in ordinary, nongrowing sheets (see different but related approaches in [3, 4, 10, 11]). The resulting effective theory has the same structure: The total energy is a surface integral of an energy density of the mid surface. The energy density is a sum of stretching and bending energies that are associated with in-plane strain and with curvature of the mid surface, respectively. In order to incorporate the known growth fields we introduce two reference 2D tensors: $\bar{a}$, the *2D reference metric*, encodes all the information about the local in-plane rest distances dictated by lateral growth. $\bar{b}$, the *reference curvature* tensor, contains the information about the local spontaneous curvature in each direction, which results from variation of growth across the thickness (these two 2D tensors replace the 3D metric tensor presented earlier). Using these reference tensors, the energy density takes the form

$$E = E_S + E_B \propto t \int (\text{Metric difference})^2 \mathrm{d}S + t^3 \int (\text{Curvature difference})^2 \mathrm{d}S$$

or

$$E\propto t \int [(1-\nu)|a-\bar{a}|^2 + \nu \mathbf{Tr}^2(a-\bar{a})]\mathrm{d}S + t^3 \int [(1-\nu)|b-\bar{b}|^2 + \nu \mathbf{Tr}^2(b-\bar{b})]\mathrm{d}S, \tag{3.3}$$

where **Tr** indicates the trace of a tensor, a and $\overline{a}$ are the actual and reference 2D metrics, b and $\overline{b}$ are the actual and reference curvatures, t is the thickness of the sheet, dS is the infinitesimal surface element, and ν is the Poisson ratio of the material. The $\propto$ relation stands for proportionality, where the coefficients are constant material parameters.

Note that in this formulation we do not use a configuration or a reference configuration. Since we deal with distances and curvatures of a 2D surface we directly use its (actual and reference) first (metric) and second (curvature) fundamental forms. This gives us simplicity, accuracy, and capability of dealing with residually stressed sheets.

Looking at (3.3) one might suggest that the system will choose a configuration in which $a = \overline{a}$ and $b = \overline{b}$ have zero elastic energy. This is indeed what happens for "regular," or *compatible*, plates and shells. However, as a and b are the forms of an actual 2D surface in space (a configuration), they are not independent. They must fulfill the Gauss–Peterson–Codazzi–Mainardi (GPCM) equations—that connect between the metric and curvature tensors of a surface [12]. For example, according to Gauss *theorem egregium*, the Gaussian curvature, which is $\det(b)/\det(a)$, is completely determined by a. In contrast, the reference tensors $\overline{a}$ and $\overline{b}$ *are* independent, not subjected to the (GPCM) equations. In fact, for general growth, they are likely to violate these equations. In this case we say that $\overline{a}$ and $\overline{b}$ are incompatible—there is no surface in R^3 for which they are the first and second forms, respectively. For the mechanical sheet it implies that there is no stress-free configuration. The body always contains elastic energy and the competition between bending and stretching leads to the selection of a and b for the given $\overline{a}$ and $\overline{b}$, i.e., shape selection.

Having an energy functional, it is possible to gain insight into the principles that govern this shape selection, to understand the global nature of the problem, and to use differential geometry for "educated guessing" of solutions in limit cases.

3.3.2 Thin/Thick Limits

What can be said about the selected configurations when we consider fixed sheet size and reference tensors, and vary the thickness t? The bending energy is cubic in t, while the stretching is linear. This implies that the thinner the sheet is, the more energetically expensive stretching is, compared to bending. Therefore, for very thin sheets the selected configurations will contain very small amount of in-plane strains. For our sheets this implies that as the thickness decreases, the actual 2D metric, a, gets closer and closer on average to the reference metric, $\overline{a}$. In the limit $t \to 0$ we will have $a \to \overline{a}$, while the curvature tensor, b, is "enslaved" to $\overline{a}$—It is optimized among all curvature tensors that are consistent with $\overline{a}$. In other terms we can say that in the limit $t \to 0$ the selected configuration is an *embedding* of $\overline{a}$ with the least bending content. This formulates a link between the physical problem and a purely geometrical problem and makes the powerful tools and theorems of differential geometry relevant for "guessing" or "understanding" the behavior of the underlying physical system. It is important to remember that this geometrical connection is valid only in the limit $t \to 0$ and only if such a limit exists (for a proof and details see [13]).

In the opposite limit—that of large thickness—bending deformations become increasingly costly compared to stretching deformations. Therefore, thick enough sheets will "obey" the reference curvature tensor having $b \sim \overline{b}$, while a is optimized among all metrics that are consistent with $\overline{b}$.

When stating that a sheet is "thin" or "thick" "enough" we need to specify with respect to what. For flat sheets of lateral scale L and thickness t, the Foppl–von-Karman number $\gamma \approx (L/t)^2$ determines the thin ($\gamma >> 1$) and thick ($\gamma \approx 1$) limits. When the reference tensors are nontrivial they define additional length scales—the inverse of reference curvatures (that are defined by both $\overline{a}$ and $\overline{b}$). When defining the thin/thick limits, one needs to take these scales into account.

3.4 Building Actively Deforming Sheets

Once the theoretical framework for studying actively deforming sheets is formulated, we turn to the nontrivial task of producing such bodies in a controllable way. The techniques that are described in this chapter are far from being optimized and most of them are suitable only for basic study of shaping principles. It is very likely that advanced techniques will be developed and will allow using active deformation as a shaping or a design technique.

3.4.1 Responsive Gels

Responsive gels are excellent materials for construction of actively deforming sheets. These gels undergo reversible swelling/shrinking transitions that are induced by external fields. Gels that respond to various fields, such as temperature [14] or concentration of chemical agents [15], are available. The activation of the gel by the external field can be sharp ("on–off-like") or gradual ("analog-like"), where the equilibrium volume is a continuous function of the stimulating field intensity. One can think of two basic approaches for building actively deforming gel sheets: In the first one the sheet composition is uniform and the driving field is designed to vary across the sheet, leading to nonuniform shrinkage/swelling that defines nontrivial reference tensors. A basic problem with this approach is how to maintain the desired local value of the driving field while the sheet changes its shape in space. In the other approach, on which we will focus, the sheet is built

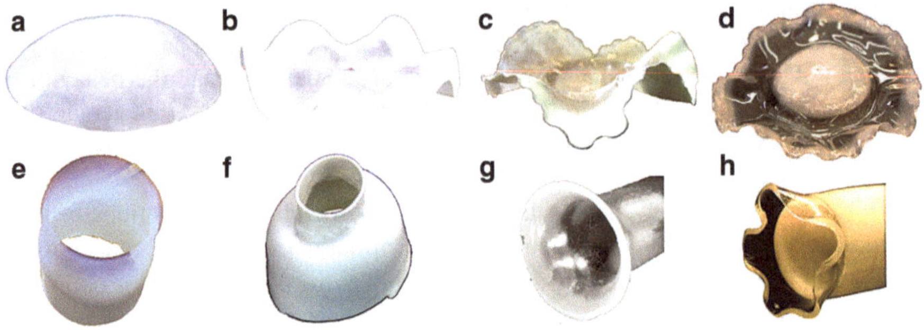

Fig. 3.2 Non-Euclidean plates and tubes made of NIPA gel. Examples of plates with $\bar{K}>0$ (**a**), $\bar{K}<0$ (**b**, **c**), and a disc that contains a central region of $\bar{K}>0$ and an outer part of $\bar{K}<0$ (**d**). (**e–h**) Non-Euclidean tubes. A tube with a "step function" metric in its cold (**e**) and warm (**f**) states. Tubes with negative curvature below (**g**) and above (**h**) the "buckling-wrinkling" transition. E. Sharon, and E. Efrati [19], reproduced with permission courtesy of The Royal Society of Chemistry

with predetermined internal gradients in its composition so that it *responses differentially to a uniform driving field.* In previous works [16–18], we used N-Isopropylacrylamide (NIPA) gel, which undergoes a sharp shrinking transition above $C^0 33$. We have found that the amount of shrinkage in the warm state is a strong function of the monomer concentration in the gel: Gels with low monomer concentration lose up to 60 % of their length at high temperature, while concentrated gels shrink by less than 10 % in length (see [19]). Once calibrating the connection between monomer concentration and shrinkage, one can start building sheets with predetermined, nontrivial reference tensors.

In order to determine some reference metric $\overline{a}$, we note that the square roots of its entries are in fact the local shrinkage values of the gel. We therefore convert the metric into a monomer concentration field across the sheet. Once activated (in a hot bath), this concentration field will lead to nonuniform, locally isotropic "attempted shrinkage" of the gel, i.e., a reference metric which is obtained by a conformal transformation of the initial flat metric. The result is a "Non-Euclidean Plate" (NEP). Examples of NEP are presented in Fig. 3.2. Since in two dimensions every metric can be obtained via a conformal transformation of a flat metric [20], one has high flexibility in determine $\overline{a}$. When designing the concentration field one should use the proper coordinate system, taking into account its variation during shrinkage.

In order to determine a reference curvature tensor, $\overline{b}$, one should introduce gradients in composition across the thickness of the sheet. Here, the isotropic shrinkage of the gel limits the accessible reference curvatures and one needs to introduce additional components into the gel in order to locally break the symmetry in the plane (see Sect. 5.2.3).

A better defined technique for metric determination is by selective UV cross-linking of a uniform polymer solution. The cross-linking density is known to be a key factor in setting the swelling/shrinkage of the gel. Therefore, the cross-linking density field takes the place of the monomer concentration field in the previous example. When it varies across the sheet, it determines a nonuniform metric. This technique provides excellent spatial resolution in determining the metric and it is likely to be developed further.

3.4.2 Other Materials

Other candidate materials for construction of actively deforming sheets are nematic elastomers. These elastic materials have an "ordered" nematic state at low temperatures and are disordered at higher temperatures [21]. Therefore, they change their metric not by locally isotropic volume changes (absorbing/expelling water), as responsive gels do, but by volume preserving transformations (changes in the order parameter). Until now, all the examples we are aware of use changes between a disordered phase and a homogeneous nematic phase that are uniform across the sheet, i.e., flat metrics. However, one can attempt to control the local orientation of the director field in the nematic phase, thus determine non-Euclidean metrics. An example of control over the reference curvature was presented in [22], where gradients across the thickness were imposed and activated. Finally, it is possible that the techniques developed to control electroactive polymers [23] could be used for metric determination.

3.4.3 Leaves as Actively Deforming Sheets

Nature uses active deformation as a common shaping mechanism at different length and timescales and via different deformation mechanisms. Examples are movement of Invertebrates or the operation of the heart. In this section we give examples of active deformation in leaves during growth. The leaf is a good system to try to model mechanically. It is a slender sheet, made of relatively rigid tissue that increases its size by many orders of magnitude during its development. Plant cells do not move with respect to each other and their cell walls form a rigid mesh, or a skeleton [24]. The description of a leaf as an actively deforming elastic sheet is, therefore, tempting.

From a mechanical point of view, the growth process determines reference metric and curvature tensors and the leaf skeleton evolves and deforms, so that the leaf is in a mechanical equilibrium at each instance. Still, the leaf is a much more complicated system than the responsive gel sheets. Its mechanical properties are nonuniform (for example, the veins are much stiffer that the rest of the tissue), the reference tensors are not known, and are affected by external signals during growth. More importantly, growth (or the reference tensors) is affected by *internal fields* within the leaf. As these fields themselves are affected by the shape of the leaf, they cause a feedback loop:

Fig. 3.3 The flexibility of leaf shape. Four leaves of Tomato mutants (**a**, **b**), wild type (**c**) and a transgene overexpressing knotted 1 gene (**d**). The relatively minor biological manipulations can prevent the development of a leaf into its natural compound shape (**c**). It can develop into simple (**a**), dissected (**b**), or ultra compound (**d**) shapes (figures were provided by N. Ori)

Growth determines reference tensors → mechanics selects a configuration → the change in configuration alters the stresses and flows within the leaf → leading to changes in the growth field (the reference tensors). The growing leaf can, thus, be viewed as a nonlinear evolving system. Under normal conditions the system is stable and growth is well regulated. However, its nonlinearity makes it capable of responding dramatically to some perturbations. Indeed, even within a single species, one can find phenotypes with highly different leaf shapes. These can result from minor "biological perturbations," such as a single gene misexpression [25, 26] or hormone application [27, 28] (Fig. 3.3). Regulation of growth is far from being clear and it is the subject of intense research. Many of the works in this field use controlled "perturbations" to the normal growth in order to expose the regulation mechanisms. These perturbations alter the reference tensors and the leaf responds by changing its 3D shape. One needs a good understanding of the connection between reference tensors and sheet configuration in order to properly analyze such data.

3.4.4 Hygroscopic Motion in Plant Tissue

Hygroscopic motion is a much simpler mechanism common in plants. This type of active deformation occurs in sclerenchymal tissue—tissue made of dead cells. It is thus a process that can be analyzed from a purely mechanical point of view, without having to consider feedback on the reference tensors. Sclerenchymal tissue typically consists of fiber cells whose walls are made of layered cellulose fibrils with preferred orientation. When absorbing/expelling water the tissue expands/shrinks anisotropically, perpendicularly to the fibrils orientation ([29, p. 200]. Changes in air humidity induce such uniaxial swelling/shrinkage that drive, for example, the

Fig. 3.4 Hygroscopic shape change. Snap shots of a Pine cone (*top*) and an Erodium awn (*bottom*) as they change configurations during a transition from a wet state (*left*) to a dry state (*right*)

opening and closure of a pine cone [30], and the penetration of wheat seeds into soil [31] (see Fig. 3.4). Seed pod opening is another example, which is reviewed in details further in this chapter.

The shape transforming structures that are made of sclerenchymal tissue are, in fact, bodies made of smart, composite material, designed to respond specifically to external conditions (changes in humidity). It is likely that such structures (that were hardly studied) can inspire biomimetic designs of actively deforming mechanical elements.

3.5 Examples

Here we give some examples of specific actively deforming structures. These examples are selected to illustrate and emphasize the principles and relevant limits that were discussed above.

3.5.1 Non-Euclidean Discs with Constant Gaussian Curvature

The first example is of NEP (see review in [19] and specific experiments in [18])—Thin sheets in which $\overline{b} = 0$ and $\overline{a} \neq I$. This example demonstrates how the configurations of NEP approach an isometric embedding of $\overline{a}$ as $t \rightarrow 0$. It also shows that the approach to an embedding can be of qualitatively different types, depending on the underlying (purely geometrical) embedding problem.

Gel discs were made by injecting the NIPA solution, through a center hole, into a gap between closely spaced glass plates. The solution is polymerized within a time of order minutes forming an elastic thin plate. NEP plates are made by a *controlled variation of the monomer concentrations*, C, during injection. As the flow field is axially symmetric, we end up with a disc in which $C = C(r)$. Such discs are flat when they are cold, but they are "programmed" to shrink by a different ratio at each radius upon heating. This differential "growth" by the shrinking ratio $\eta = \eta(r) = \eta(C(r))$ prescribes a new target metric (in polar coordinates) of the form

$$\bar{a} = \begin{pmatrix} \eta^2 & 0 \\ 0 & r^2\eta^2 \end{pmatrix}.$$

Introducing the arc-length radial coordinate on the shrunk disc ρ,

$$\rho(r) = \int_0^r \eta(r')\mathrm{d}r',$$

transforms the target metric to

$$\bar{a}(\rho) = \begin{pmatrix} 1 & 0 \\ 0 & \eta^2(\rho)\rho^2 \end{pmatrix}, \tag{3.4}$$

for which the imposed Gaussian curvature is

$$\overline{K} = -\frac{1}{\eta\rho}\frac{\partial^2(\eta\rho)}{\partial\rho^2} \tag{3.5}$$

In order to construct *discs of constant Gaussian curvature* whose target Gaussian is constant, we solve (3.5) for $\overline{K} = \mathrm{Const}$ and obtain the reference metrics:

$$\bar{a} = \begin{pmatrix} 1 & 0 \\ 0 & \frac{1}{\overline{K}}\sin^2\sqrt{\overline{K}}\rho \end{pmatrix}$$

for $\overline{K}>0$, and

$$\bar{a} = \begin{pmatrix} 1 & 0 \\ 0 & \frac{1}{-\overline{K}}\sinh^2\sqrt{-\overline{K}}\rho \end{pmatrix}$$

for $\overline{K}<0$.

We, then, tune the monomer concentration field to produce the calculated functions above and produce the gel discs. In the experiments that are described below, we keep the disc radius fixed $R = 28$ mm and set the reference Gaussian curvature to $\overline{K} = \pm 0.0011\ \mathrm{mm}^{-2}$. The initial thickness of the disc was the only parameter that was changed ($0.1\ \mathrm{mm}<t<1.5\ \mathrm{mm}$).

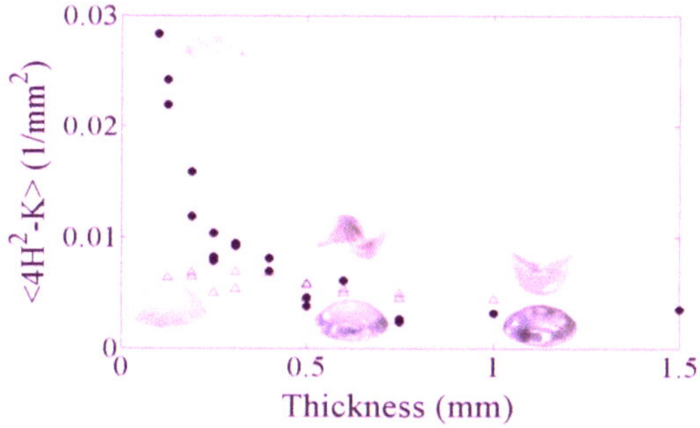

Fig. 3.5 The average bending content as a function of sheet thickness. For discs of $\bar{K}<0$ (*solid circles*) the bending content increases sharply with decreasing thickness to values an order of magnitude larger than $|\bar{K}|$. For $\bar{K}_r>0$ the bending content is "saturated" at small thickness. *Insets* show selected discs at the relevant thickness. Discs diameter is 28 mm and $|\bar{K}| = 0.0011$ mm

3.5.1.1 Results

For $\overline{K}>0$ the discs buckled into dome-like shapes, preserving the axial symmetry of the reference metric (Fig. 3.5, bottom insets). Such configurations were predicted theoretically for the thin limit. In this limit the theoretical analysis shows that the deviation from the isometric embedding (a perfect spherical dome) was in the form of a *boundary layer* whose width scales like $t^{1/2}$. In this boundary layer the curvature perpendicular to the edge vanishes. This condition is often "put by hand" as a boundary condition when using local equations of force and torque balance. In our formalism it is derived directly from energy minimization (see [32]). The experimental data is consistent with these calculations. All discs assumed dome-like configurations with Gaussian curvature very close to $\overline{K}$. The predicted boundary layer is visible and it shrinks as thickness decreases (Fig. 3.5, bottom insets). Finally, the bending energy of these discs is cubic in the thickness. This scaling is expected, since as $t \rightarrow 0$ the system converges to the embedding of $\overline{a}$ of least bending (in this case it is a spherical dome of radius $1/\sqrt{\overline{K}}$) and nearly all the change in the bending energy is due to the factor t^3 in (3.3). The thick limit was studied theoretically [32]. It was found that very thick discs are perfectly flat (obeying $\overline{b}$) down to a critical buckling threshold thickness t^*. Below t^* there is a second-order transition and the disc buckles with increasing amplitude as the thickness further decreases.

The situation is qualitatively different for discs with $\overline{K}<0$. Though on average these discs nearly obey the reference metric, they do not settle on axially symmetric configurations. Instead, they break the axial symmetry and form wavy configurations (Fig. 3.5, top insets). Thick sheets adopt a saddle shape, but as the thickness decreases these configurations are replaced by wavy ones. The number of waves scales like $t^{-1/2}$. A direct consequence of the observed refinement is a slower decrease of the bending energy as t decreases. The bending content is roughly quadratic in the number of waves, leading to the scaling $E_b \sim t^2$, which is similar to energy scaling in some wrinkling problems. This refinement and slow decay of energy are surprising and not fully understood. Several theoretical works have addressed this topic, but the observation is not fully explained yet. It is important to note that the only difference between the two cases is the sign of the reference Gaussian curvature. All length scales and symmetries are the same. The origin of the qualitative difference between the cases of $\overline{K}<0$ and $\overline{K}>0$ should thus be in the embedding problem of the reference metrics.

3.5.2 A Strip with Imposed Hyperbolic Curvature Tensor

The previous example considered the case in which $\overline{b} = 0$ and $\overline{a}$ is non-Euclidean—prescribing nonzero Gaussian curvature. Now we will look at the opposite case: A sheet for which $\overline{a}$ is trivial, i.e., the identity matrix, and $\overline{b} \neq 0$. In particular, we will study long strips with a reference curvature that describes a sheet that locally "wants" to bend into a saddle shape at every point. The principal directions of $\overline{b}$ are oriented at some angle θ from the long axis of the strip.

It was recently discovered that this type of structure is common in many plant organs. In particular, the microstructure of many seed pod valves determines such reference tensors, utilizing the underlying mechanics to drive hygroscopic opening of the seed pod. We therefore review in parallel the biological and synthetic systems.

3.5.2.1 The Opening of a Seed Pod

The dynamic opening of a pod, leading to seed dispersal, is another example of shape transformation driven by active deformation. This process is called dehiscence and it is common in *Leguminosae*. Most Legumes blossom in spring and their seeds are developed during early summer. At this stage, the pod is made of living cells, and its valves are flat or convex, keeping the pod closed, protecting the seeds. When the seeds are fully developed, the pods' cells die, left to dry in the summer air. At the right time the pods crack open, its two valves are separated, spreading the seeds. The opening can be abrupt, shooting the seeds to large distances, or slow. *Bauhinia variegata* pods dehisce explosively by a rapid curling of their valves into oppositely twisted helical strips (Fig. 3.6 left). The flat-to-helical transition occurs when the valve tissue is dead (sclerenchymal tissue, see Sect. 4.4). It is, therefore, not a biological process, but a mechanical one. What is the mechanism converting the initially flat valves into oppositely curled helical strips, and how is it related to the microscopic structure of the pod tissue?

Bauhinia valves are elongated narrow strips with typical length, width, and thickness of $L \approx 15$ cm, $w \approx 2$ cm, and $t \approx 1$ mm, respectively. They twist when

Fig. 3.6 The opening of a seed pod. As they dry up (*left top to bottom*) the two initially flat valves gradually twist in two opposite handedness, driving the opening of the pod. Cutting narrow strips in $\pm 45^\circ$ to the pod axis and drying them leads to their bending in opposite directions without twisting. The $\pm 45^\circ$ directions are, thus, the principal directions of the reference curvature tensor and are defined as the x and y directions

free of external constraints, driven by the active deformation of the tissue [33]. The valve is made of fiber cells whose walls are made of layered cellulose fibrils with a preferred orientation. When absorbing/expelling water the tissue expands/shrinks anisotropically, perpendicularly to the fibril orientation (see Sect. 4.3).

In Bauhinia, the pod valves consist of two fibrous layers, oriented roughly at $\pm 45^\circ$ with respect to the pod's longitudinal axis. How can this simple architecture lead to the observed shape transition?

In order to answer this question, we determine the reference metric and curvature of the pod: It is well known that a uniaxial shrinkage of one layer determines a curvature along the shrinkage direction, where the shrunk (shorter) layer is in the inside of the curved surface. When two layers are shrunk in perpendicular directions, each prescribes a curvature. The result is two opposite curvatures in two perpendicular directions—a saddle shape. We take the curvatures to be equal, and use the principal directions (the $\pm 45^\circ$) to write the reference curvature:

$$\bar{b} = \begin{pmatrix} k_0 & 0 \\ 0 & -k_0 \end{pmatrix}, \tag{3.6}$$

Since the valve material is laterally uniform, the reference metric is flat and can be written as

$$\bar{a} = \begin{pmatrix} 1 & 0 \\ 0 & 1 \end{pmatrix} \tag{3.7}$$

To verify that these are indeed the reference tensors, one can cut narrow strips from a wet valve at $\pm 45^\circ$. When drying, these strips bend into arcs in two opposite directions (Fig. 3.6, right).

Fig. 3.7 A small disc (*left*) and a large rectangle (*right*) that were cut from the same composed latex sheet having reference curvature and metric as in (3.6) and (3.7). The small disc is governed by $\bar{b}$, adopting a *saddle shape*. The wide sheet is governed by $\bar{a}$ adopting a *cylindrical shape*

We also construct a synthetic model material, which is made by stretching two latex sheets in perpendicular directions (the x- and y-directions) and gluing with flexible glue. When free of external constraints, the composed sheet shrinks actively in two perpendicular directions along its top and bottom layers, similarly to the pod tissue. This active deformation prescribes a reference curvature tensor as in (3.6), with k_0 determined by the initial stretching of the sheets (see [34]).

Once the reference tensors are determined, we are ready to discuss the shaping principle of the valve. The incompatibility of this type of sheet is clear: The flatness (Zero Gaussian curvature) of $\overline{a}$ is inconsistent with the negative Gaussian curvature $\overline{K} = -k_0^2$ that is prescribed by $\overline{b}$. Therefore, the bending and stretching terms cannot be minimized simultaneously by any configuration. They have to compete. As explained earlier, $\overline{b}$ will be dominant in the "thick limit," while $\overline{a}$ will be obeyed in the thin limit. Cutting small and large circles from the sheet demonstrates the different limit behavior. Small disc attains a saddle shape (Fig. 3.7), as determined by $\overline{b}$. The large disc rolls into a cylindrical shape—a configuration of zero Gaussian curvature that "satisfies" one of the principal reference curvatures. Indeed, these cylindrical configurations are bi-stable and can flip between configurations that obey k_0 or $-k_0$. In these configurations a bending-dominated boundary layer is visible. Unlike the case of the NEP, where the boundary layer was developable (since in these sheets $\overline{b} = 0$), here the boundary layer is saddle-like, obeying $\overline{b}$.

3.5.2.2 Strips

We now turn to apply the above principles to the case of long strips geometry. The thin/thick limits can be identified as follows: A flat strip (a configuration that obeys $\overline{a}$) will have bending energy that scales like $E_b \sim t^3 w k_0^2$. A strip configuration that obeys $\overline{b}$ (a minimal surface) will have stretching energy that scales like

$E_s \sim tw(k_0 w)^4$. The ratio between the two energies varies like $w^4(k_0/t)^2$. This is the relevant dimensionless number that takes the role of the Föppl–von-Karman number in flat sheets. One sees that stretching is dominant for large width and imposed curvature and for small thickness, while bending is dominant in the opposite limit. Using a dimensionless width, $\tilde{w} = w\sqrt{k_0/t}$, we expect stretching dominance for $\tilde{w}>>1$, and bending dominance for $\tilde{w}<<1$.

<u>The $\tilde{w}<<1$ limit:</u> Here we have $b \sim \overline{b}$. The shape is determined by the reference curvature, while the metric is "enslaved." We therefore expect the strips to adopt configurations of minimal surfaces that coincide with $\overline{b}$. For the valve, in which the principal directions of $\overline{b}$ are in $\pm 45°$ with respect to the pod axis, these configurations are in the form of a "twisted ribbon" with a well-defined handedness. Helices of opposite handedness are generated from the same material if the orientation of the pod axis (with respect to the principal directions) is changed by 90°. This is exactly the difference between the two pod valves, leading to curling in opposite handedness, assuring opening of the pod.

<u>The $\tilde{w}>>1$ limit:</u> Here we have $a \approx \overline{a} = 1$. The configurations must be developable surfaces—surfaces with zero Gaussian curvature, and $b - \overline{b}$ is optimized among these. It can be shown that the selected configurations are "helical ribbons" that are "cut from a cylinder." The cylinder has radius $1/k_0$ (up to corrections due to Poisson's ratio) and the pitch angle of the helix is simply θ, the angle between the principal directions and the pod axis. For $\theta \neq 45°$ these configurations are bi-stable and can be flipped to obey either k_0 or $-k_0$, as demonstrated in Fig. 3.8c. Similarly to the thin limit, a change of 90° of θ leads to the same configuration in an opposite handedness.

3.5.2.3 Experimental Verification

The two calculated limits above can be compared to experimental data from both valve tissue and a synthetic structure. Synthetic sheets are made from two sheets of Latex that are stretched along perpendicular directions and glued together. The result is a composed sheet with reference curvature and metric as in (3.6) and (3.7) (see details in [34]).

To study the flat-to-helical transition, we cut from the composed sheet strips of different width, w, and angle, θ, and study their shape. Cutting at $\pm 45°$ leads to strips with geometry similar to the pod valve, but other angles are equally interesting. The strips attend different helical configurations that are characterized by the pitch and radius of their center line, and by the distribution of mean and Gaussian curvatures of the surface. The calculated helical shapes are in excellent agreement with experimental data. For $\theta = 45°$ we observe the transition, in both valve tissue and synthetic sheets, from "twisted ribbons" ($r = 0$) at $\tilde{w}<<1$ to "helical ribbons" at $\tilde{w}>>1$ (Fig. 3.8a). The transition between the two families of solutions occurs at $\tilde{w} \approx 4$ (Fig. 3.8b). Surface measurements on synthetic strips confirm that the

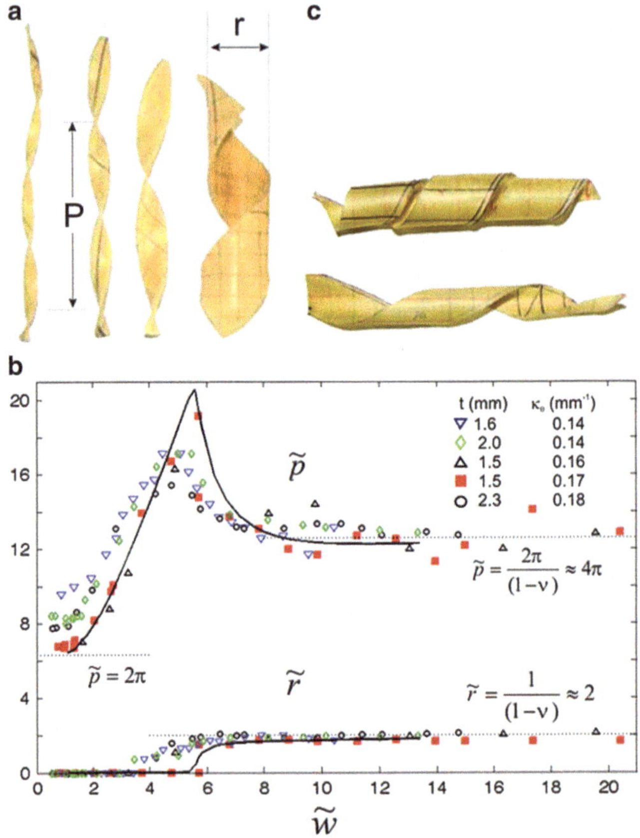

Fig. 3.8 Helical configurations and transitions. (**a**) Strips cut at 45° from the same composed latex sheet with $\bar{a}$ and $\bar{b}$ as in (3.6) and (3.7). As the width of the strip increases (*left to right*) a twisted-to-helical transition is observed. (**b**) Measurements of the normalized pitch and radius, $\tilde{p}$ and $\tilde{r}$, of helical latex strips ($\theta = 45°$) as functions of $\tilde{w}$. The thickness and spontaneous curvatures are indicated in the figure. *Dashed lines* indicate the calculated values of $\tilde{p}$ and $\tilde{r}$ in the narrow and wide limits. The *solid lines* are intermediate values that were obtained numerically. (**c**) Two stable configurations of a wide strip cut at 60°. In each of the two configurations the sheet bends along a different principal curvature

transition is from minimal surfaces at $\tilde{w} << 1$ to developable surfaces at $\tilde{w} >> 1$. In the second limit a bending dominated boundary layer is clearly detected and the helices are bi-stable (Fig. 3.8c). Finally, the measured variation of pitch and radius with θ is also in excellent agreement with the calculated values [34].

It is important to note that $\tilde{w}$ is dimensionless and its increase can result from different processes: an actual increase in width, an increase in curvature, or a decrease in thickness. This is demonstrated a NIPA gel model of the valve. The samples are prepared by casting homogeneous NIPA gel and immersing in it two perpendicular sets of parallel threads close to its top and bottom surfaces. Since the gel is prevented from shrinking/swelling parallel to the threads, its active deformation is uniaxial,

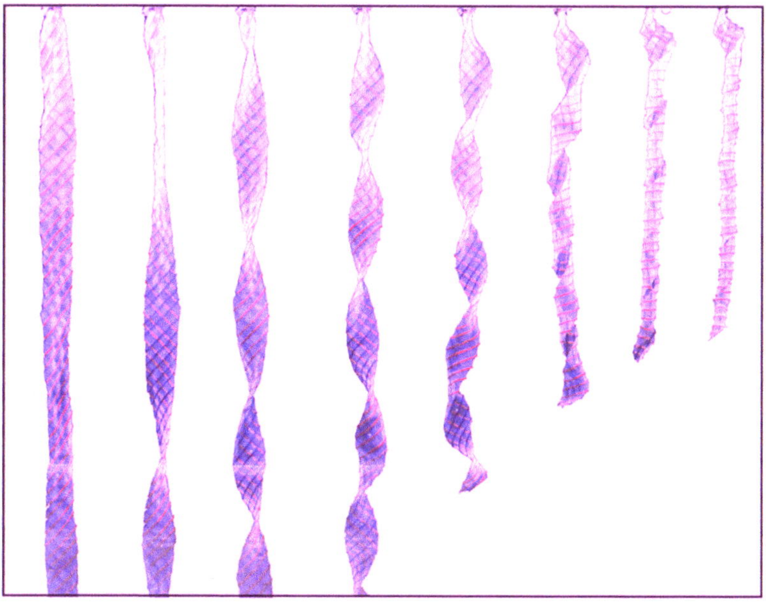

Fig. 3.9 Flat-twisted-helical-tube transition. A NIPA gel strip, reinforced with *blue* and *red cotton threads* (see text) as it heated from room ~30 (*left*) to 45 °C (*right*). The initially flat strip undergoes conformation transitions: The flat configuration is replaced by a twisted one. The pitch of the twisted configurations decreases, until a helical configuration replaces the twisted one. On further heating, the pitch of the helix decreases until the strip touches itself, forming a tube. Qualitatively similar transitions have been observed during self assembly of macromolecules from chiral elements

perpendicular to the thread direction. This leads to a sheet that grows/shrink in two perpendicular directions along its two faces, i.e., a sheet with reference curvature as in (3.6). We cut strips from the composed gel sheet and put them in water bath. When increasing the temperature of the water the gel increasingly shrinks, leading to a gradual increase in k_0 and a decrease in t. This leads to a total increase in $\tilde{w}$. Indeed, a reversible flat–twist–helical transition is observed in this system (Fig. 3.9).

3.5.2.4 The Connection to Shape Selection of Helical Macromolecules

The inputs to the energy functional are geometrical quantities—the reference metric and curvature tensors. The theoretical formulation does not include information about fine scale structure. This is not a coincidence—Amorphous elastic materials are perfectly described by 3D reference and actual metrics. Within 2D modeling, the state of an amorphous sheet is fully described by reference and actual 2D metric and curvature tensors. Therefore, many different microscale architectures can lead to the same functional. For a given microstructure of a sheet, one only needs to find $\bar{a}$ and $\bar{b}$ and then the trends and limit behavior of the structure become clear and solvable.

For example, the same $\bar{b}$ as in (3.6) can result even if the layers are not perpendicular to each other, or even when there is only one layer of fiber tissue on top of an isotropic layer (see [34]). Indeed, the thorough study of pods from over 300 species of Leguminosae by Fahn and Zohary [35] revealed a vast variety of microscale architectures. Some of these architectures drive pod opening as in the Bauhinia and some drive different mechanical transitions. These architectures can be analyzed using the modeling above and inspire the construction of smart active structures.

Surprisingly, the mechanism studied here is highly relevant to systems in a completely different scale and context. These are macromolecules that are generated via self-assembly of chiral molecules. Such macromolecules often attain configurations in the form of helical strips [36, 37] and some of them undergo a "twisted-to-helical ribbon" transition during growth [38]. Some models of these monolayers assume that the chirality of the molecules introduces a term in the bending energy of the sheet, which accounts for an *intrinsic twist* (see [39–41]) along the long axis of the strip. These models predict the observed transition. It turns out that such a system is *mechanically equivalent to the pod valve system.* To see that we go back to the reference curvature of the valve, which takes the form

$$\bar{b} = \begin{pmatrix} k_0 & 0 \\ 0 & -k_0 \end{pmatrix}$$

when written in a coordinate system aligned with its principal directions ($\pm 45°$ with respect to the pod axis). If we use a coordinate system aligned with the pod axis (we rotate coordinate system by 45°) we find

$$\bar{b} = \begin{pmatrix} 0 & k_0 \\ k_0 & 0 \end{pmatrix} \tag{3.8}$$

This reference curvature tensor represents a sheet whose only nontrivial property is a spontaneous twist along its long axis. This is exactly the mechanical characteristics of the macromolecules described above. The mechanical behavior of such structures that was studied above is, thus, relevant to shape selection of such macromolecules. Another system that was recently studied is a sheet made of nematic elastomers with intrinsic twist along the z-direction [22]. The system undergoes a similar twist to helical transition as the amount of twist increases. One can easily show that the microstructure of these strips leads to reference curvature as in (3.6) and (3.8). In this system, it is the increase in k_0 that caused the increase in $\tilde{w}$ and drives the flat–twist–helical transition. The examples above show that shape formation and transitions observed in chemical systems and composed materials might be explained by the effective elastic formulation presented in this chapter. It seems that the Bauhinia pod had opened an exciting direction of study, combining biology, chemistry, physics, and engineering.

3.6 Summary

The fascinating physics of actively growing elastic bodies is in the early stages of being formulated and revealed. This young branch of mechanics contains many open questions and challenges and is naturally connected to diverse scientific fields: The contribution of mathematicians is essential for understanding the underlying geometrical problems (the embedding problem) and the type of energy minimizers. Biologists can both integrate the new mechanical concepts into morphogenetical models and discover new "smart" natural active structures that can affect the further development of the field. Experimental and theoretical physicists are expected to revile the basic shaping principles and to develop experimental techniques for the construction of active bodies. Chemists and material scientists can both contribute to the design of new materials and the application of the new concepts to molecular structures. Finally, engineers and even industrial designers can take the accumulated basic knowledge and turn it into functioning devices, such as "soft machines."

References

1. Truesdell C (1952) The mechanical foundations of elasticity and fluid dynamics. Indiana Univ Math J 1:125
2. Landau LD, Lifshitz EM (1986) Theory of elasticity. Pergamon, Oxford
3. Efrati E, Sharon E, Kupferman R (2009) Elastic theory of unconstrained non-Euclidean plates. J Mech Phys Solids 57:762
4. Yavari A (2010) A geometric theory of growth mechanics. J Nonlinear Sci 20:781
5. O'neill B (1997) Elementary differential geometry. Academic, New York
6. Ciarlet PG (1997) Theory of plates, vol 27. North-Holland Publishing Co., Amsterdam
7. Ciarlet PG (2000) Theory of shells, vol 27. North-Holland Publishing Co., Amsterdam
8. Pogorelov AV (1988) Bendings of surfaces and stability of shells. American Mathematical Society, Providence
9. Dervaux J, Ciarletta P, Ben Amar M (2009) Morphogenesis of thin hyperelastic plates: A constitutive theory of biological growth in the Föppl-von Kármán limit. J Mech Phys Solids 57:458
10. Lewicka M, Mahadevan L, Pakzad MR (2011) The Foppl-von Karman equations for plates with incompatible strains. Proc Math Phys Eng Sci 467:402
11. Struik DJ (1961) Lectures on classical differential geometry. Dover, New York
12. Lewicka M, Pakzad MR (2009) Scaling laws for non-Euclidean plates and the W2;2 isometric immersions of Riemannian metrics. arXiv:0907.1857 [math-ph]
13. Tanaka T (1978) Collapse of gels and critical endpoint. Phys Rev Lett 40:820
14. Philippova OE et al (1997) pH-responsive gels of hydrophobically modified poly(acrylic acid). Macromolecules 30:8278
15. Efrati E et al (2007) Spontaneous buckling of elastic sheets with a prescribed non-Euclidean metric. Phys Nonlinear Phenom 235:29
16. Klein Y, Efrati E, Sharon E (2007) Shaping of elastic sheets by prescription of non-Euclidean metrics. Science 315:1116
17. Klein Y, Venkataramani S, Sharon E (2011) Experimental study of shape transitions and energy scaling in thin non-Euclidean plates. Phys Rev Lett 106(11):118303

18. Sharon E, Efrati E (2010) The mechanics of non-Euclidean plates. Soft Matter 6:5693
19. Warner M, Terentjev EM (2003) Liquid crystal elastomers. Oxford University press, New York
20. Sawa Y et al (2011) Shape selection of twist-nematic-elastomer ribbons. Proc Natl Acad Sci U S A 108:6364
21. Bar-Cohen Y (2002) Electroactive polymers as artificial muscles: a review. J Spacecraft Rockets 39:822
22. Taiz L, Zeiger E (2002) Plant physiology, 3rd edn. Sinauer Associates, Sunderland, MA, p I
23. Nath U et al (2003) Genetic control of surface curvature. Science 299:1404
24. Ori N, Eshed Y, Chuck G, Bowman JL, Hake S (2000) Mechanisms that control knox gene expression in the Arabidopsis shoot. Development 127:5523
25. Fleming AJ (2002) The mechanism of leaf morphogenesis. Planta 216:17
26. Sharon E, Marder M, Swinney HL (2004) Leaves, flowers and garbage bags: making waves. Am Sci 92:254
27. Fahn A, Werker E (1972) Anatomical mechanisms of seed dispersal. In: Seed biology. New YOrk, NY: Academic Press
28. Reyssat E, Mahadevan L (2009) Hygromorphs: from pine cones to biomimetic bilayers. J R Soc Interface 6:951
29. Elbaum R et al (2007) The role of wheat awns in the seed dispersal unit. Science 316:884
30. Efrati E, Sharon E, Kupferman R (2009) Buckling transition and boundary layer in non-Euclidean plates. Phys Rev E 80:016602
31. Armon S et al (2011) Geometry and mechanics in the opening of chiral seed pod. Science 333:1726
32. Fahn A, Zohary M (1955) On the pericarpial structure of the legumen, its evolution and relation to dehiscence. Phytomorph 5:99
33. Oda R et al (1999) Tuning bilayer twist using chiral counterions. Nature 399:566
34. Zastavker YV et al (1999) Self-assembly of helical ribbons. Proc Natl Acad Sci U S A 96:7883
35. Ziserman L et al. (2011) Curvature instability in a chiral amphiphile self-assembly. Phys Rev Lett 106
36. Ghafouri R, Bruinsma R (2005) Helicoid to spiral ribbon transition. Phys Rev Lett 94:4
37. Selinger JV, Spector MS, Schnur JM (2001) Theory of self-assembled tubules and helical ribbons. J Phys Chem B 105:7157
38. Selinger RLB et al (2004) Shape selection in chiral self-assembly. Phys Rev Lett 93:4

Chapter 4
Ion Beam-Induced Self-Assembled Wrinkles

Myoung-Woon Moon, Chansoo Kim, and Ashkan Vaziri

Abstract Instability of a stiff thin film attached to a compliant substrate generally results in the appearance of exquisite wrinkles with length scales that depend on the system geometry and applied stresses. Several methods have been developed for creating surface wrinkles including inducing compressive stresses/strains on a thin metal deposited on a polymer substrate, dewetting polymer, and UVO/ion beam-irradiated polymeric surface.

In this work, we have reviewed the formation of ion beam-induced self-assembled wrinkle patterns on polymer surfaces. Exposure to ion beam generally results in formation of a stiff skin on surface areas of a polymeric surface. The created stiff skin has strain mismatch with the polymeric surface, leading to generation of ordered surface wrinkles. By controlling the ion beam fluence and area of exposure of the poly(dimethylsiloxane) (PDMS), one can create a variety of patterns in the wavelengths in the micron to submicron range, from simple one-dimensional wrinkles to peculiar and complex hierarchical nested wrinkles. The induced strains in the stiff skin can be estimated by measuring the surface length in the buckled state. The patterned surfaces have a variety of cross-disciplinary applications that range from optics and electronics to tissue engineering and regenerative medicine. One novel usage of these patterns is for fabricating wrinkles with extreme topology. As an example, by using the prefabricated wrinkle pattern by ion beam, we developed wrinkles with high aspect ratio of amplitude over wavelength. Here, first the wrinkles were induced on a PDMS surface using Ar

---

M.-W. Moon (✉) • C. Kim
Future Convergence Research Division, Korea Institute of Science and Technology (KIST), Seoul, Republic of Korea
e-mail: mwmoon@kist.re.kr

A. Vaziri
Department of Mechanical and Industrial Engineering, Northeastern University, Boston, MA 02115, USA

X. Chen (ed.), *Mechanical Self-Assembly: Science and Applications*,
DOI 10.1007/978-1-4614-4562-3_4,

ion beam irradiation. The wrinkles had a wavelength in the range of 200–1,400 nm depending on the ion treatment time. Then, an amorphous carbon film was deposited on the pre-patterned PDMS to elevate the amplitude of surface features using a glancing angle deposition.

4.1 Introduction

Instabilities of thin films (or skins) that generally lead to large out-of-plane deformations (e.g., wrinkling) and possible failure and cracking, impose significant challenges to many different industries such as flexible electronics, cell templates, nanochannels for protein condensation, or other fields [6, 8, 16]. Especially residually compressed thin films on thick substrates may delaminate and buckle on hard substrate as energy release rate exceeds the interface fracture toughness [15, 20, 21]. While films or skins could be compressed on compliance substrate, it may wrinkle along with deformation of substrate [6, 9, 23]. Wrinkle configurations range from straight shape to herringbone and hierarchical with respect to the stress distribution and level [22]. Under uniaxial compression applied externally, a thin film (or a stiff layer) attached to a compliant substrate forms a straight wrinkle patterns with well-defined sinusoidal shape, which might be used for cell templates, nanochannel of protein condensation, or optical grating devices. These systems could be maintained under equi-biaxial stress state in pre-deposited thin film as shown in Fig. 4.1a or ion beam-induced wrinkled skin on polymer in Fig. 4.1b [22, 23], inducing herringbone or more complex wrinkle patterns or further hierarchical patterns with respect to stress level and thickness of stiff layer.

In case of a compressed stressed film deposited on a soft polymer, creating nonlinear wrinkle patterns were introduced as an alternative method for lithography patterning [26]. A nanoscale wrinkle pattern was formed by depositing a thin compressed diamond-like carbon (DLC) film, which causes wrinkling due to its internal residual stress of 0.5 GPa as well as large difference in Young's modulus between thin DLC film (~100 GPa) and poly(dimethylsiloxane) (PDMS) substrate (~2 MPa).

It has been reported that the high compressive stress in a DLC film is originated from the knock-on implantation of carbon atoms with an optimum ion energy to overwhelm the stress relaxation during the deposition process [26]. Furthermore, a significant amount of sp^3 bonds would also contribute to the high compressive stress of the DLC film since the sp^3 sites tend to form a local compression stress rather than tension in the sp^2 sites. Especially, unlike other methods, typical external forces such as pre-stretching and/or thermal contraction of the substrate are not required to create wrinkles on a compliant substrate. Figure 4.1a shows the formation and evolution of the wrinkled surface with single (left)- and double (right)-wavy modes as a function of film thickness. The wavelength of a single-wavy mode proportionally increases with the DLC film thickness. At a larger film thickness, the wavelength deviates from the theory, presumably due to coarsening

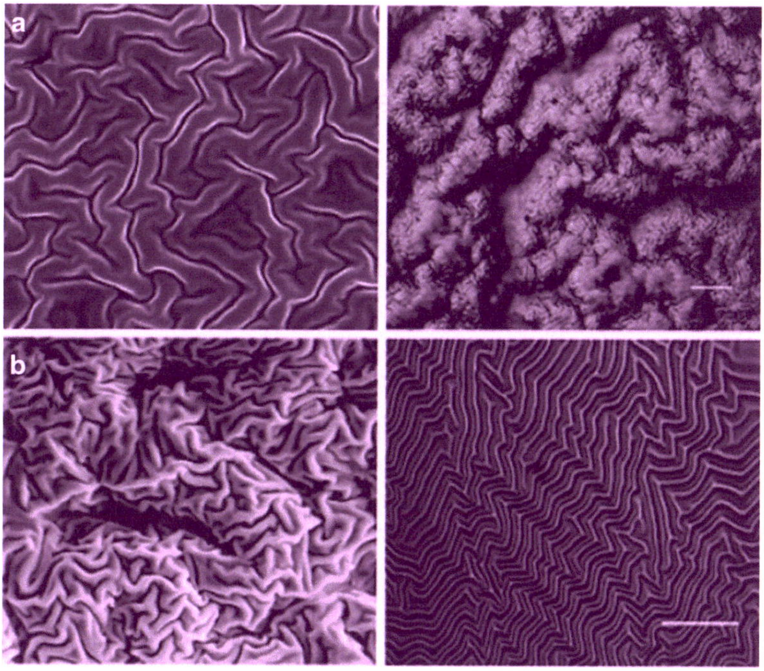

Fig. 4.1 Images for wrinkle patterns induced on a polymeric surface by deposition of a thin compressive film (**a**) and by ion beam irradiation or etching process (**b**)

of the neighboring waves. Hierarchical wrinkles with two distinguishable wavelength of approximately 100 nm and 2,000 nm were observed at a higher film thickness [26].

Without deposition of new materials on soft polymers, the attempt to develop surface patterns for polymers has yielded various methods from lithography and plasma and UVO treatment to liquid polymer films dewetting, and thin metal film deposition on polymers [3, 10, 11, 13, 30]. These techniques have been employed to create a variety of patterns, such as dots, network structures, complex hierarchically assembled patterns, honeycomb-like structures, and ringlike patterns on the surface of polymers.

Among various methods for fabrication of wrinkle patterns, surface modification by ion bombardment or plasma treatment has been previously investigated for glassy metals, as well as amorphous and crystalline materials [4, 12]. Generally, ion beam irradiation normal to the surface leads to surface milling—as also seen in our recent experiments on polyimide, a polymer much stiffer than PDMS [24]. In these experiments, the ion effect is characterized by the mean penetration depth and longitudinal and lateral straggling widths, depending on material properties. On the other hand, ion beam irradiation at non-normal incident angle leads to appearance of intricate and often complex surface features. For polyimide, the structural features manifest in the form of ripples.

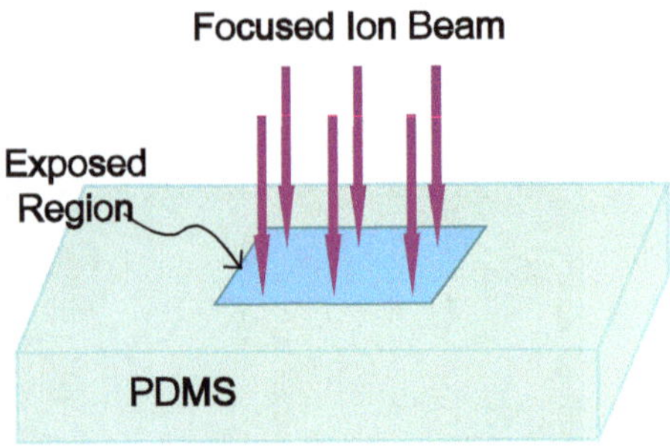

Fig. 4.2 A schematic of wrinkle patterns on PDMS surface induced by FIB irradiation

In recent experiments on PDMS, it has been shown that focused or broad ion beam irradiation can be used for creation of self-assembled wrinkle patterns on the polymeric surfaces [22, 23, 25]. The flexibility provided by this technique provides new avenues for creation of structural features at micron and submicron scales on the surface of polymers.

As an example, irradiation of Ga ions beam (FIB) on the surface of a PDMS substrate, Fig. 4.2, results in formation of a stiff skin on the substrate. This stiff skin experiences in-plane compressive strain upon formation induced by ion beam [23] and buckles to accommodate the induced mismatch strain between the skin and the polymeric substrate. This finding provides a robust technique for creating wrinkling patterns on selective areas of PDMS by simply controlling the movement of the ion beam relative to the polymeric substrate. A Ga ions beam with a wide range of acceleration voltages and ion fluencies created various pattern configurations from straight one-dimensional undulations to herringbone and hierarchical patterns with multiple wavelengths. The role of the acceleration voltage and the fluence of the ion beam are explored by systematically varying these parameters between 5 and 30 keV and 10^{12} to 10^{18} ions/cm^2, respectively. To extend the capabilities of this technique, maskless patterning was developed for selecting the location of the surface patterns. This method allows the accurate selection of the wrinkling areas exposed to the FIB, confined by Bitmap images of the exposure patterns. Moreover, we will discuss the fabrication of wrinkles with high amplitude/wavelength ratio by the deposition of an amorphous carbon film on a surface of a pre-patterned soft polymer PDMS. In the experiments, glancing angle deposition (GLAD) was used to deposit amorphous carbon film on a PDMS surface.

4.2 Theoretical Background

Basic theory for wrinkling instability of stiff skin or film layers on soft substrates under compression has been studied comprehensively [3, 7]. Wrinkling geometries of wavelength and amplitude and onset strain are determined by the combination of film thickness h and the ratio of elastic moduli between film (E_f) and substrate (E_s).

By considering the wrinkling geometries under plane strain condition on infinitely deep substrate, wrinkling configuration could be assumed as $w = w_{\max} \sin(\pi x/L)$, and a critical strain for the onset of wrinkle would be expressed by

$$\varepsilon_c = 1/4\left(3\overline{E}_{\text{sub}}/\overline{E}_f\right)^{2/3}, \tag{4.1}$$

where $\overline{E} = E/(1 - v^2)$. Above the critical strain, ε_c, for stiff layer, the wavelength (L) of wrinkling would be predicted as

$$L/h = \alpha\sqrt[3]{\overline{E}_f/\overline{E}_{\text{sub}}}, \tag{4.2}$$

here α is given about 4.36 for plane strain condition [7]. Noting that wavelength L would increase by increasing the thickness h_f of film and the ratio of elastic modulus, $\overline{E}_f/\overline{E}_{\text{sub}}$. The amplitude of the sinusoidal wrinkling patterns, A, can be predicted from

$$A/h = \sqrt{(\varepsilon/\varepsilon_c)^2 - 1}, \tag{4.3}$$

where ε is the compressive strain applied to the stiff skin.

4.3 Ion Beam-Induced Surface Patterns on Polymers Using Ion Beam

The precise mechanisms underlying formation of these wrinkling patterns have not been completely understood. However, it was shown that ion beam irradiation creates formation of a stiff skin on the surface of PDMS, as also seen in previous experiments on the effect of ion beam irradiation on metallic surfaces. It was analyzed that this thin stiff skin resembles amorphous silica due to selective depletion of carbon by ion [23, 27]. The thin stiff skin induced on the soft (or rubberlike) polymeric surface undergoes in-plane compression upon formation and buckles leading to creation of wrinkles. As ion beam irradiation on various materials was used for high-precision micro- and nanomachining, it has been discussed that ion beams would induce compositional and morphological changes because of ion–solid interaction resulting in ion implantation or polymer chain modification. As the ion beam creates or induces surface defects or polymer chain modification, it has been reported that the biaxial compressive stress will develop [17].

In our experiments, wrinkle patterns are formed by exposing the surface area of a flat PDMS sheet (thickness ~ 3 mm, Young modulus $\approx$ 2 MPa) to an FIB of Ga ions as schematically shown in Fig. 4.3. This method can create wrinkle patterns of various widths and complexity by controlling the relative motion of the polymeric

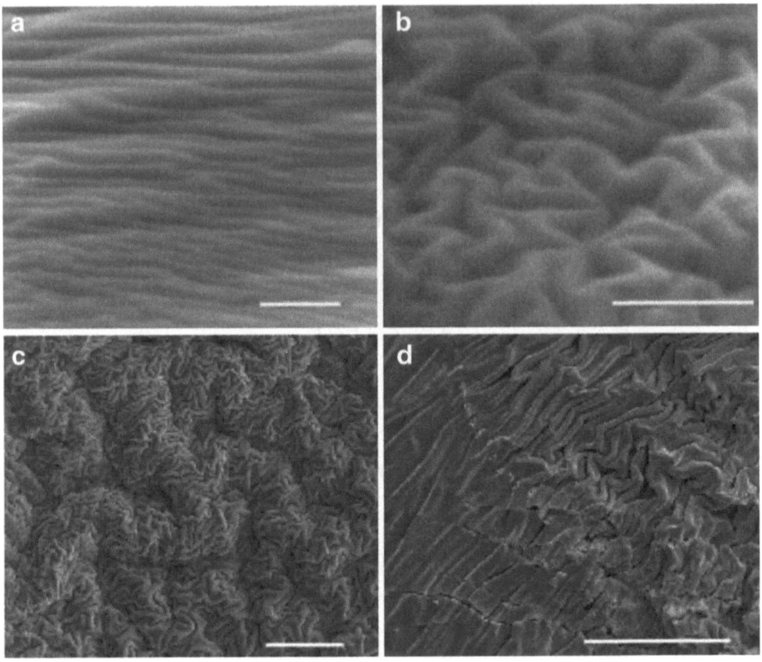

Fig. 4.3 Selected SEM images of the wrinkling patterns. Panels (**a**, **b**) display one-dimensional straight buckles and herringbone patterns, respectively. (**c**) Complex hierarchical patterns with primary wavelength ~465 nm (Bar = 5 μm). (**d**) Complex wrinkling patterns and surface cracking were observed (Bar = 5 μm)

substrate and the FIB to scan selected areas as discussed below. Because of the strain relaxation to release ion beam energy accumulated in the stiff skin, the wrinkles appear only on the exposed areas of the PDMS to FIB. FIB exposure creates a tendency for the skin to expand in the direction perpendicular to the direction of FIB irradiation, if it were not constrained by the PDMS substrate. The mismatch strain between the newly formed stiff skin and its substrate gives rise to skin instability and formation of the wrinkles. FIB exposure differs from UVO treatment of PDMS in that the latter produces a stiff skin by increasing cross-links with relatively little strain mismatch. The morphology of the wrinkle patterns on the surface areas of PDMS is mainly a function of ion fluence. In our experiments, the incident angle defined as the angle between the incoming beam and the surface normal was maintained at 0° and the beam dwell time was 3 μs in all cases.

The wavelength and morphology of the induced undulations can be effectively selected by controlling the accelerating voltage and fluence or current density of the ion beam as shown in Fig. 4.3 [23]. Figure 4.3a and b displays one-dimensional straight buckles and herringbone patterns, respectively, with primary wavelength ~50 nm created using acceleration voltage 5 keV and ion fluences 9.0×10^{13} ions /cm^2 and 5.0×10^{14} ions/cm^2, respectively. Nonlinear wrinkles with hierarchical wavelength were evolved with primary wavelength ~465 nm created using an acceleration voltage 30 keV and ion fluence 1.0×10^{16} ions/cm^2. The wavelength was almost linearly dependent on the accelerating voltage or ion beam energy. As increased the ion fluence up to 1.0×10^{17} ions/cm^2 for an acceleration voltage 30 keV, complex wrinkling patterns, and surface cracking were observed in Fig. 4.3d.

Various morphologies shown in Fig. 4.4a were created by a single-mode FIB scanning with the beam current of 1 nA and the fluence level denoted in the images. When the PDMS substrate is exposed to ion beam at a fluence in the order of 1×10^{13} ions/cm^2, induced wrinkles are mainly straight and one dimensional with an average wavelength $\approx$ 460 nm. Herringbone wrinkles form at a fluence of 5×10^{13} ions/cm^2. Nested hierarchical patterns are created at a fluence of about 7×10^{13} ions/cm^2, or greater, with primary wrinkles having average wavelength of $\approx$ 460 nm nested in the larger secondary wrinkles with average wavelength ≈ 2.0 μm, revealing the hierarchical nature of wrinkling structures.

Another method for controlling the morphology of the wrinkles is multiple scanning of the same region by an FIB with constant beam current. Figure 4.4a shows SEM images of the surface morphology of the PDMS exposed to FIB in a single scan mode ($N = 1$) with various ion beam fluences, F (ions/cm^2). Figure 4.4b shows SEM images of the surface morphology created by N scans at a constant fluence, $F = 2.0 \times 10^{13}$ ions/cm^2. SEM images of the surface after $N = 15$ and $N = 20$ scans reveal complex hierarchical patterns. By subjecting the surface to multiple scans at the same acceleration voltage, we verified that the morphology of the wrinkling patterns is indeed controlled by the accumulative ion fluence, defined simply as the number of scans times the ion fluence per scan.

4.3.1 Energy Dependency

To quantify the morphology of the wrinkle patterns, we examined the topology of the wrinkles using atomic force microscopy (AFM) in the tapping mode (Fig. 4.5a). For the fluences of approximately 10^{13} ions/cm^2, a periodic, one-dimensional profile appears on the regions exposed to FIB, while at fluences below this level, the surface of the PDMS remains flat. Figure 4.5b shows the average induced strain in the stiff skin as a function of FIB fluence for the acceleration voltages 10, 20, and 30 keV, respectively. The compressive strain in the stiff skin induced by FIB irradiation was estimated by direct measurement of the surface length, L, along a trace across the surface. With L_0 as the straight-line distance between the ends of the trace, the strain approximation is taken as $(L - L_0)/L_0$. The lowest ion fluence which causes appearance of one-dimensional straight buckles is in the order of 10^{13} ions/cm^2 with a slight dependence on the acceleration voltage. The average induced strain at the onset of skin wrinkling is $\varepsilon_c \sim 3\,\%$ for the three sets of measurement shown in Fig. 4.5b. Examination of the wrinkling patterns created by ion beam with

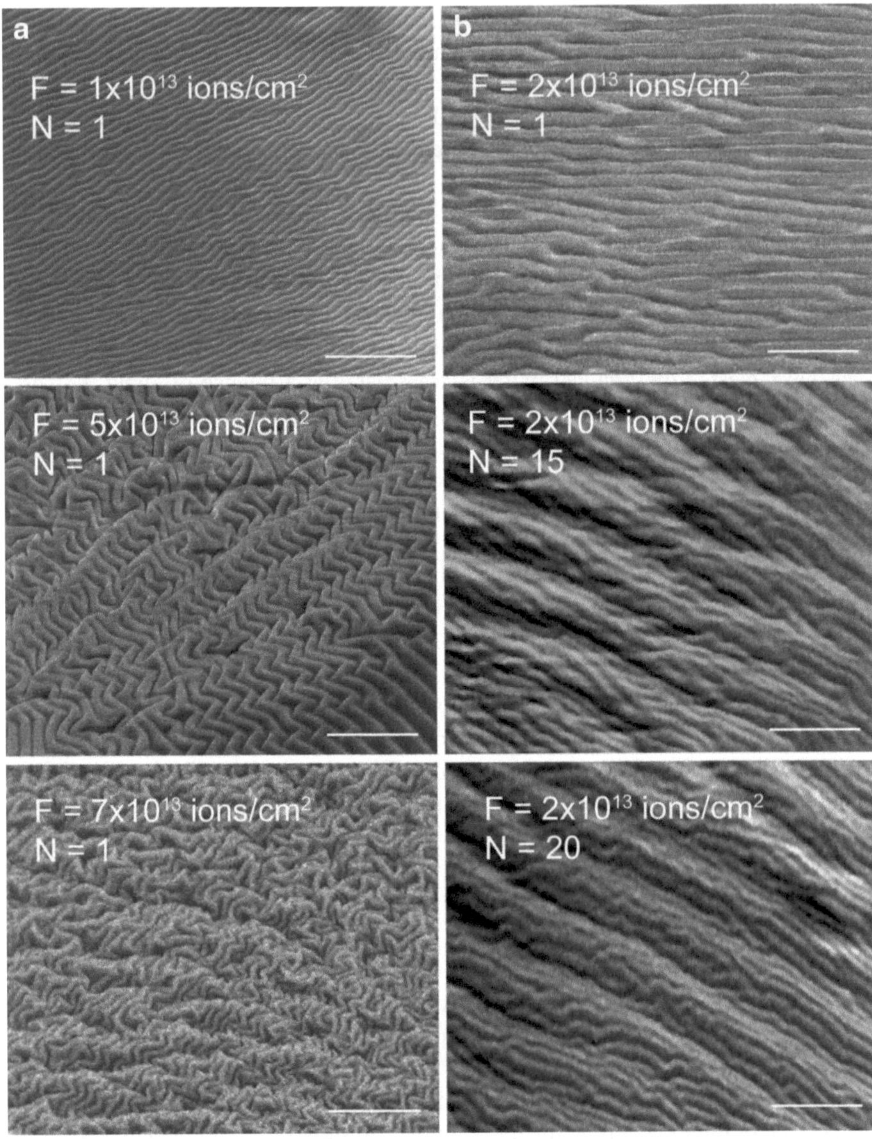

Fig. 4.4 The surface morphology of the wrinkle patterns induced by FIB depends primarily on the applied ion fluence. (**a**) SEM images of the surface morphology of the PDMS exposed to FIB in a single scan mode ($N = 1$) with various ion beam fluences, F (ions/cm^2). (**b**) SEM images of the surface morphology created by N scans each with fluence, $F = 2.0 \times 10^{13}$ ions/cm^2. Scale bars = 10 μm. Moon et al. [23], copyright 2007 National Academy of Science, USA, reprinted with permission

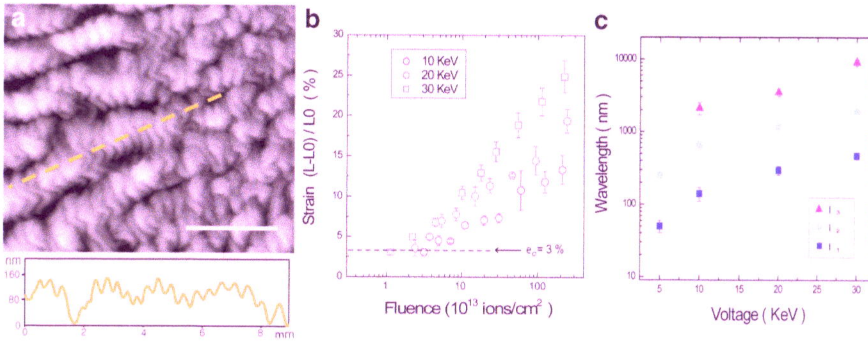

Fig. 4.5 Examination of the surface topology and chemical composition of the PDMS exposed to FIB through depth. (**a**) An AFM image and cross-sectional profile, (**b**) average compressive strain in the stiff skin as a function of ion fluence for acceleration voltages, and (**c**) wavelengths of the wrinkling patterns. Moon et al. [22], © Carl Hanser Verlag, Muenchen, reprinted with permission

acceleration voltage of 10 and 20 keV confirmed that the induced average strain in the skin at the onset of wrinkling formation is effectively independent of the ion beam acceleration voltage. The classical relationship for buckling of a linear elastic stiff skin with modulus, E_s, attached to a compliant substrate with elastic modulus, E_f, gives the critical strain associated with the onset of instability as (4.1), $\varepsilon_c \approx 0.52(E_s/E_f)^{2/3}$, independent of the skin thickness. Based on $\varepsilon_c \sim 3\,\%$, the modulus ratio is from (4.2), $E_f/E_s \approx 70$. The associated wavelength, λ_1, of the first wrinkles to form, referred as the primary wrinkles, scales with the thickness of the stiff skin, t, according to $\lambda_1/t \cong 4(E_f/E_s)^{1/3}$.

Close examination of the undulations also shows that the wavelengths of the patterns depend primarily on the acceleration voltage. A critical ion fluence is required to produce a given pattern, but the fluence has little effect on the wavelength once the pattern has formed. These observations are consistent with the notion that the acceleration voltage sets the depth of penetration of the ions and therefore the thickness of the stiff skin, while the lateral strain induced by the FIB is controlled by the fluence. The three wavelengths plotted as a function of acceleration voltage in Fig. 4.5c are measured within the hierarchal regime (see Fig. 4.6). The finest wrinkling pattern has $\lambda_1 \approx 50$ nm and was created with an acceleration voltage 5 keV, while the wrinkling patterns induced by an acceleration voltage 30 keV have $\lambda_1 \approx 450$ nm. The largest measured wavelength is $\lambda_3 \approx 10$ μm for a hierarchical pattern induced by an acceleration voltage 30 keV.

4.3.2 *Wrinkling Map*

To further investigate the relationship between the pattern morphology and the ion beam parameters, the PDMS surface was exposed to FIB of Ga+ ions in a digital

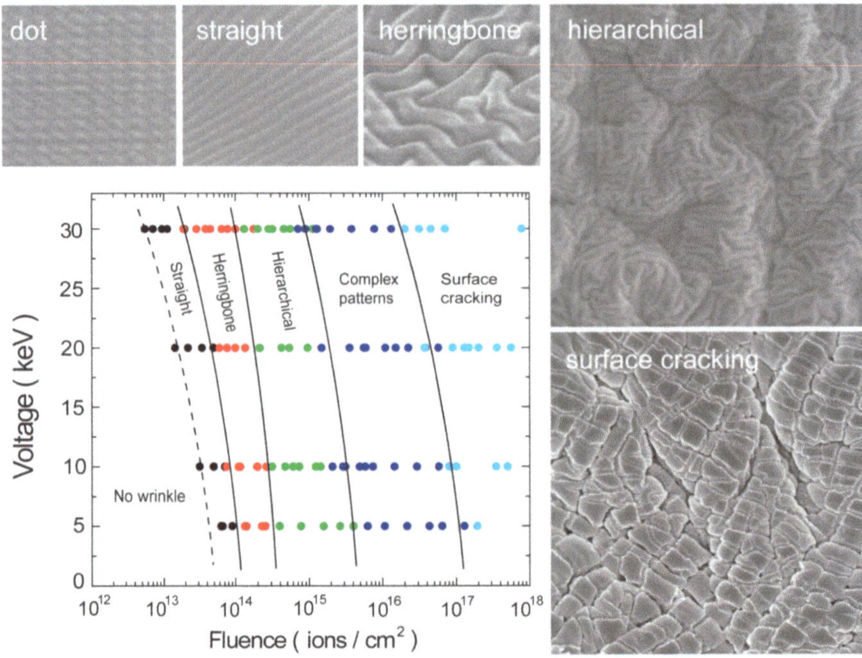

Fig. 4.6 Morphology of the wrinkling patterns created by varying the ion beam acceleration voltage and fluence. Map of wrinkle morphology as a function of FIB acceleration voltage and ion beam fluence

mode with multiple acceleration voltages and ion current in the range of 1 pA to 20 nA [22]. The general conditions for the sample preparation and ion beam were the same as those in previous sections. Figure 4.6 shows the mapping between the morphology of the wrinkling patterns and acceleration voltage and ion fluence of FIB. At each acceleration voltage, straight, one-dimensional buckles appear at low ion beam fluence, while at higher ion fluence complex patterns develop including herringbone and double-scale morphologies with two or more distinct wavelengths as seen in Fig. 4.6. The critical value of ion fluence associated with the onset of appearance of a given wrinkle pattern is higher at lower acceleration voltages. At very high fluence complex patterns and cracking of the surface were observed, denoted by "surface cracking" region.

To extend the capabilities of this technique, we have developed three different methods to select the location of the surface patterning as shown in Fig. 4.7. The area exposed to the ion beam can be selected by controlling the relative movement of the ion beam and polymeric substrate, or alternatively for applications that need precise control over the exposed area by the maskless patterning method. A significant advantage of the surface modification offered by the technique discussed here is that wrinkles appear only on the areas of the PDMS exposed to the FIB. Areas covered by wrinkles can be selected by simply controlling the motion of the ion beam relative to

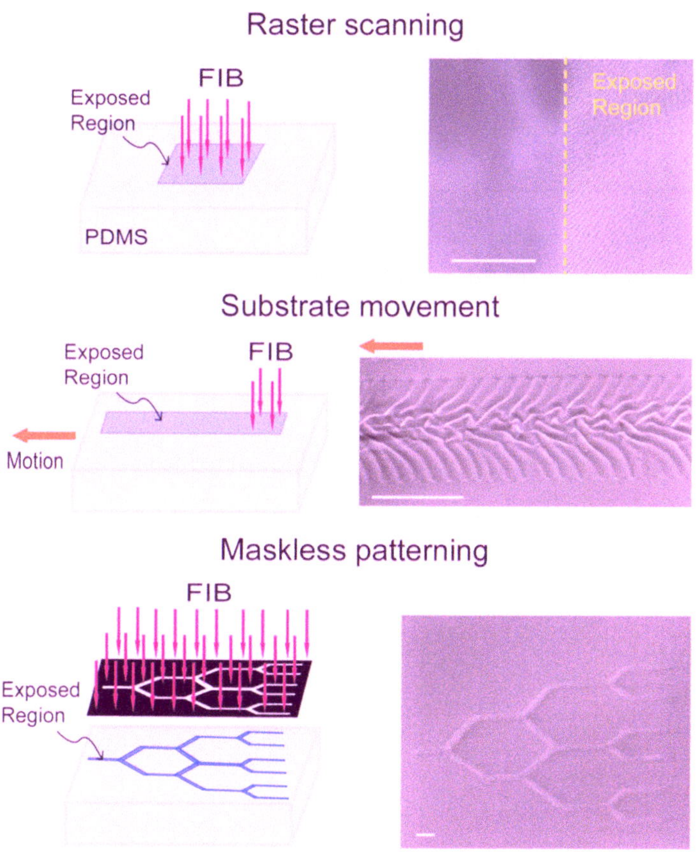

Fig. 4.7 Surface modification of polymers by focused ion beam. The figure displays the three methods developed for controlled wrinkle patterning on desired surface areas of the polymer. Scale bars = 10 μm. Moon et al. [25], reprinted with permission

the substrate (Fig. 4.7a) [23]. Here, the PDMS substrate is moved at a constant speed relative to the FIB (Fig. 4.7b). The wrinkle patterns shown in Fig. 4.7b are formed by moving the PDMS at a constant speed of 500 nm/s while the FIB fluence is controlled by decreasing the width of the exposed area from 50 μm to 4 μm at a constant beam current of 1 pA. The path of the wrinkle patterns can be selected simply by controlling the relative motion of the substrate and ion beam, while the morphology of the surface can be changed by controlling the ion fluence. This method permits the formation of paths of rough wrinkles on the surface of the PDMS that might be useful in developing multifunctional microfluidic devices [22].

Application of the maskless patterning method permits accurate selection of the areas exposed to the FIB. Bitmap files of the exposure patterns, such as those shown in Fig. 4.8, were imported as a virtual mask in the FIB system. The patterning was controlled with the ion fluence in the range of 1.3×10^{15} to 2.1×10^{16} ions/cm^2. In Fig. 4.8, maskless patterning method was used to study the role of ion fluence on

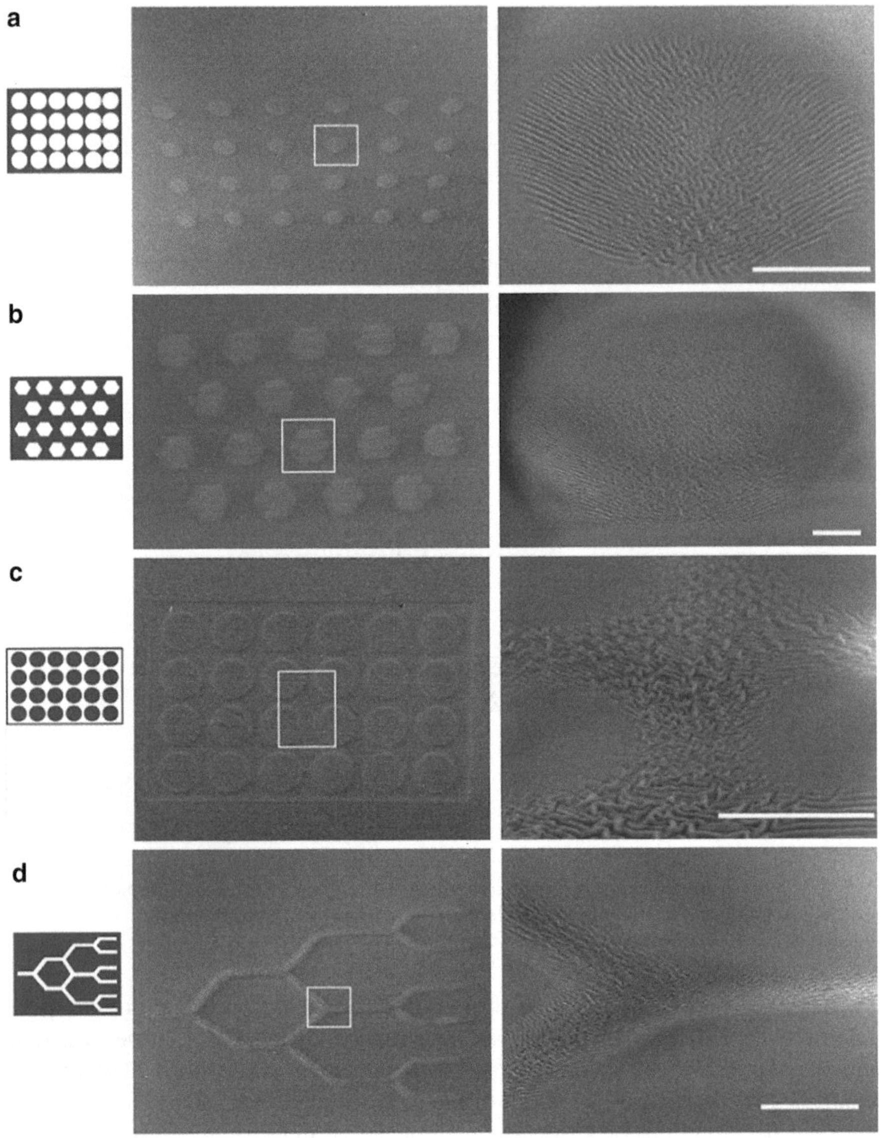

Fig. 4.8 Selective patterning of the PDMS surface using maskless patterning. The bitmap files (shown on the left for each pattern) were imported to the FIB such that only the white regions were exposed. The ion fluence of the FIB within the each patterned shape was 1.3×10^{15}, 2.1×10^{16}, 2.25×10^{15}, and 2.3×10^{15} ions/cm^2 for (**a**), (**b**), (**c**), and (**d**), respectively. Scale bar = 5 μm. Moon et al. [22], © Carl Hanser Verlag, Muenchen, reprinted with permission

creation of wrinkle patterns by FIB. The bitmap files (shown on the left for each pattern) were imported to the FIB such that only the white regions were exposed. Using a low-energy ion beam of acceleration voltage, 10 keV, wrinkling patterns with wavelength ~120 nm and amplitude of 5–30 nm were created on the exposed regions of the substrate. The ion fluence of the FIB within the each patterned shape was 1.3×10^{15}, 2.1×10^{16}, 2.25×10^{15}, and 2.3×10^{15} ions/cm^2 for (a), (b), (c), and (d), respectively.

4.3.3 Dwell Time Effect

In literatures, pixel dwell time of ion beam is considered as one of the key factors that determine the accuracy of surface modification by ion beam [25]. The dwell time controls the ion beam broadening and ion beam shifting as shown before for experiments on semiconductors [14]. In Fig. 4.9, we explore the role of ion beam dwell time on the wrinkle patterns created on the surface of PDMS using FIB irradiation. In this set of experiment, the accelerating voltage and ion current were 30 kV and 50 pA, respectively. In the experiments, flat polymeric substrates were subject to ion beam irradiation with various dwell times, while the formation of wrinkling patterns was monitored as a function of irradiation duration and therefore ion fluence. Maskless patterning was used to create the wrinkling patterns on predefined complex regions of the polymeric substrate.

The SEM figures display the wrinkle patterns created after 2 s of FIB irradiation, or equivalently at the ion fluence of $\sim 1.85 \times 10^{13}$ ions/cm^2. The dwell time was 0.5 μs and 5 μs in Fig. 4.9b and c, respectively. For the dwell time of 0.5 μs, straight wrinkles appear at the edge of the exposed region which evolves to herringbone patterns with wavelength of ~ 350 μm by moving towards the center of the exposed region. For the dwell time of 5 μs, well-defined herringbone wrinkles were fabricated on the entire area exposed to the ion beam. In Fig. 4.9a, the boundaries of the region exposed to ion beam are not very distinct due to ion beam shifting during the repetitive scanning. However, by increasing the dwell time, this effect diminishes as a longer dwell time leads to less number of repetitive scanning at constant fluence. Long-term ion beam irradiation may lead to overflow of ion dose or ion shifting outside the patterned region due to increase in the positive ion charging effect on the surface. This effect is well known for nonconductive materials, especially polymers, subject to ion beam irradiation [5].

4.4 High-Aspect-Ratio Wrinkle

As discussed earlier, the wrinkle wavelength is determined by the system geometry (mainly the film thickness) and the relative stiffness of the film and substrate, and is in general much larger than the film thickness and much smaller than the

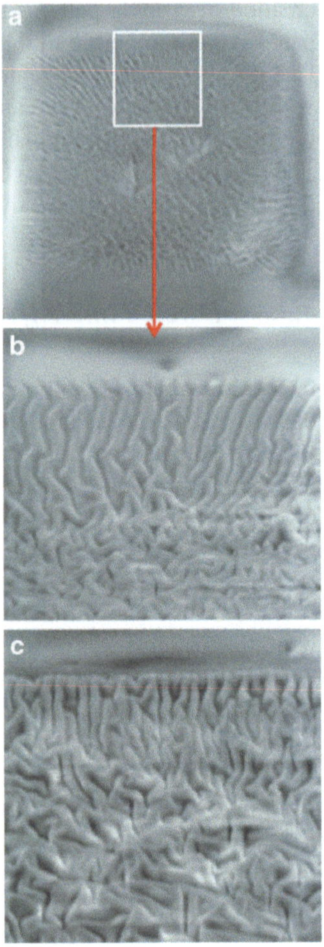

Fig. 4.9 Role of pixel dwell time on the morphology of patterns created by focused ion beam. (**a**, **b**) Wrinkle pattern created at the pixel dwell time of 0.5 μs. (**c**) The edge of the region exposed to focused ion beam with the pixel dwell time of 5 μs. Moon et al. [25], reprinted with permission

specimen dimensions. The amplitude of the wrinkles depends on the applied stresses and the magnitude of the induced strain mismatch. The ratio of wrinkle amplitude/wavelength is normally limited to 1/10 [19, 23]. Here, we use carbon film deposition on patterned surfaces to achieve high-aspect-ratio wrinkle morphologies.

As the first step, we employed GLAD for deposition of an amorphous carbon film on a flat PDMS surfaces. Amorphous carbon films are used as a protective layer in structural systems and biomedical components due to their low friction coefficient, wear resistance, and high elastic modulus and hardness [26]. The deposited carbon layer is generally under high residual compressive stresses (~1 GPa), making it susceptible to buckle delamination on a hard substrate (e.g., silicon or glass) [20] and to wrinkling on a soft substrate [23]. GLAD is a deposition method used to fabricate functional thin films with a columnar morphology.

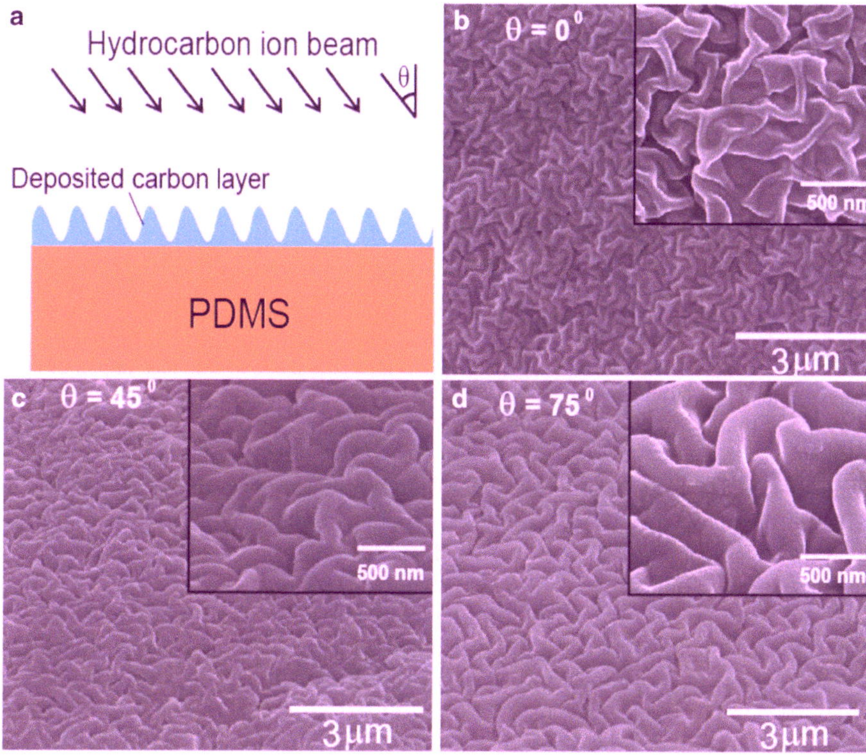

Fig. 4.10 Surface patterns created by carbon film deposition. (**a**) Schematic of the experiments, (**b–d**) scanning electron microscopy (SEM) images of the PDMS surface after 50-min carbon film deposition at different incident angles. Ahmed et al. [1], reprinted with permission

Figure 4.10a shows the schematic of a PDMS substrate subjected to hydrocarbon ion beam irradiation, and the SEM images of the wrinkles created by 50-min amorphous carbon film deposition at three different incident angles [1]. The deposited carbon film is under approximately equi-biaxial compressive stress, and the created wrinkles are semi-herringbone or semi-labyrinth shapes, Fig. 4.10b–d. The wavelength of the wrinkles is relatively insensitive to the deposition angle and duration and is ~ 750 nm. In contrast, the amplitude of the wrinkles depends on the deposition angle and duration. The wrinkles created by 50-min carbon deposition normal to the substrate surface (i.e., $\theta = 0^{\circ}$, Fig. 4.10b) has an average amplitude 144 nm (i.e., amplitude/wavelength ~1/5) and has the appearance of nonlinear wrinkle configurations observed in a biaxially compressed film on a compliant substrate. The patterns formed by deposition at 45° and 75° have approximately the same wavelength, but much higher amplitudes (amplitude/wavelength ratios of ~2 and 2.5, respectively). Figure 4.11a–c displays the SEM images of surface patterns created by carbon film deposition at 75° with different deposition durations, showing that the wrinkle amplitude increases for longer deposition durations. Ion

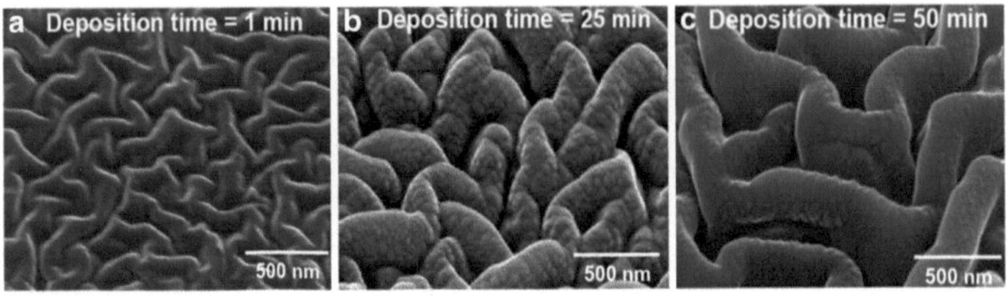

Fig. 4.11 The effect of deposition angle and time on the morphology. (**a–c**) SEM images of the PDMS surface after carbon film deposition at 75° incident angles for three different deposition durations. Ahmed et al. [1], reproduced by permission of The Royal Society of Chemistry. The article in which this figure was originally published is located at the following link: http://pubs.rsc.org/en/content/articlelanding/2010/sm/c0sm00386g

or plasma treatment at an oblique angle results in the formation of a porous thin film with anisotropic features that are induced by the atomic-scale shadowing, or self-shadowing [18, 28, 29]. As the ion or radical nucleates on the target substrate, the deposition rate is higher in front of the nucleus than the deposition rate behind it due to the shadowing by the nucleus, resulting in a porous film structure. At early stages of the deposition, wrinkles could emerge due to the relaxation of the strain energy in a compressively stressed thin film as shown in Fig. 4.11a. For a longer deposition time, the porous structure keeps growing on the wrinkled surface, resulting in an increase in the surface amplitude as shown in Fig. 4.2b and c. It was reported that the amplitude/wavelength ratio of the patterns created with the incident angle 75° increases by increasing the deposition duration, resulting in an amplitude/wavelength ratio as high as 2.5.

As an extension of this technique, we demonstrated that the amorphous carbon deposition on a pre-patterned polymeric surface allows fabrication of high-aspect-ratio wrinkles. Figure 4.12 shows the schematic of our experiments; first, we pre-patterned the surface of PDMS by Ar ion beam irradiation (Fig. 4.12a, b). The created patterns are in the form of two-dimensional wrinkles with a wavelength that depends on the treatment time and was varied between 200 and 1,400 nm in this study. In the next step, an amorphous carbon film gets deposited on the pre-patterned surface using GLAD to elevate the amplitude of the patterns (Fig. 4.12c). The details of the proposed method were outlined in the ref. [1], where we demonstrated an interesting application of the created high-aspect-ratio wrinkles for controlling the optical band gap with respect to the wavelength of wrinkles.

Wrinkle patterns with a high amplitude/wavelength ratio were fabricated with irradiation of Ar ion beam on a PDMS substrate and a subsequent deposition of an amorphous carbon film using GLAD. PDMS substrates were prepared by a mixture of elastomer and cross-linker in a mass ratio of 10:1 (Sylgard-184, Dow Corning, MI, USA). The mixture was placed in a plastic box and stirred to remove trapped air bubbles and then cured at 80 °C for 2 h, resulting in a cross-linked PDMS network, which was cut as coupons of 20 mm × 20 mm × 3 mm for the experiments.

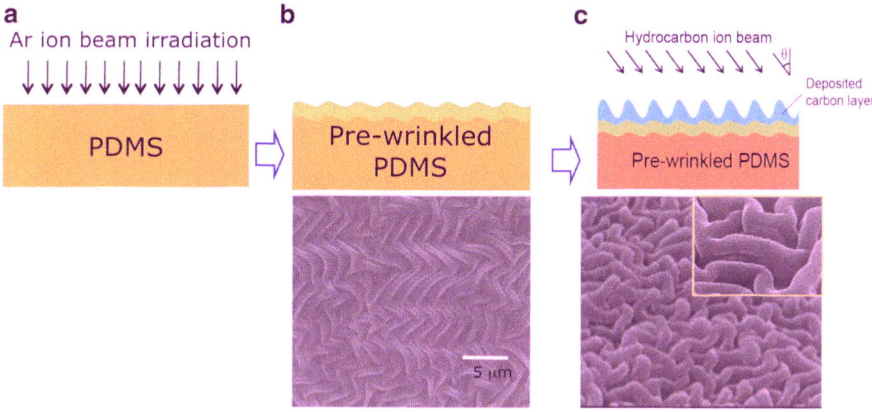

Fig. 4.12 Schematic of the fabrication of pre-wrinkle patterns on PDMS by Ar ion beam (**a**, **b**) and subsequent carbon deposition using GLAD to increase the amplitude of the pre-patterned surface (**c**). Representative SEM images for each condition. Ahmed et al. [1], reproduced by permission of The Royal Society of Chemistry. The article in which this figure was originally published is located at the following link: http://pubs.rsc.org/en/content/articlelanding/2010/sm/c0sm00386g

The Ar ion beam treatment of PDMS and also the carbon film deposition were carried out in a linear ion gun (DC 3 kV/ 6 kW, EN Technologies) [2]. The sample coupons were placed in the ion beam chamber, and the chamber was evacuated to a base pressure 2×10^{-5} mbar. The PDMS was exposed to Ar ion beam for 10 s to 50 min, leading to creation of wrinkles. The amorphous carbon film was deposited on flat as well as pre-patterned PDMS coupons by introducing the acetylene (C_2H_2) into the ion gun at a flow rate of 8 sccm. During carbon deposition, the anode voltage was kept at a constant value of 1 kV and a radio frequency (r.f.) bias voltage was applied to the substrate holder at a bias voltage of −200 V. In this study, the carbon deposition time was kept between 30 s and 50 min, while the incident angle of hydrocarbon ion was varied from 0° to 75°. Detailed conditions were referred to ref. [1].

Figure 4.13 shows AFM profile images of three different wrinkle patterns created on a PDMS surface by Ar ion beam irradiation. Ion beam irradiation results in the formation of a thin stiff skin on the polymer surface which is ~100 times stiffer than PDMS and is under compressive stress [23]. The morphology of the created patterns depends on the state of stress in the thin film. For the ion beam irradiation normal to the polymeric surface, the state of stress in the thin film is semi-equal biaxial and the wrinkles are semi-labyrinth shape. The wrinkle wavelength mainly depends on the film thickness, t, and the ratio of elastic moduli of thin film and substrate, E_f/E_s, and can be estimated from (4.2).

Figure 4.14 shows SEM images of four different patterns created on pre-patterned PDMS surfaces. The amorphous carbon deposition angle and duration were 75° and 10 min, respectively, while the duration of Ar ion beam irradiation in

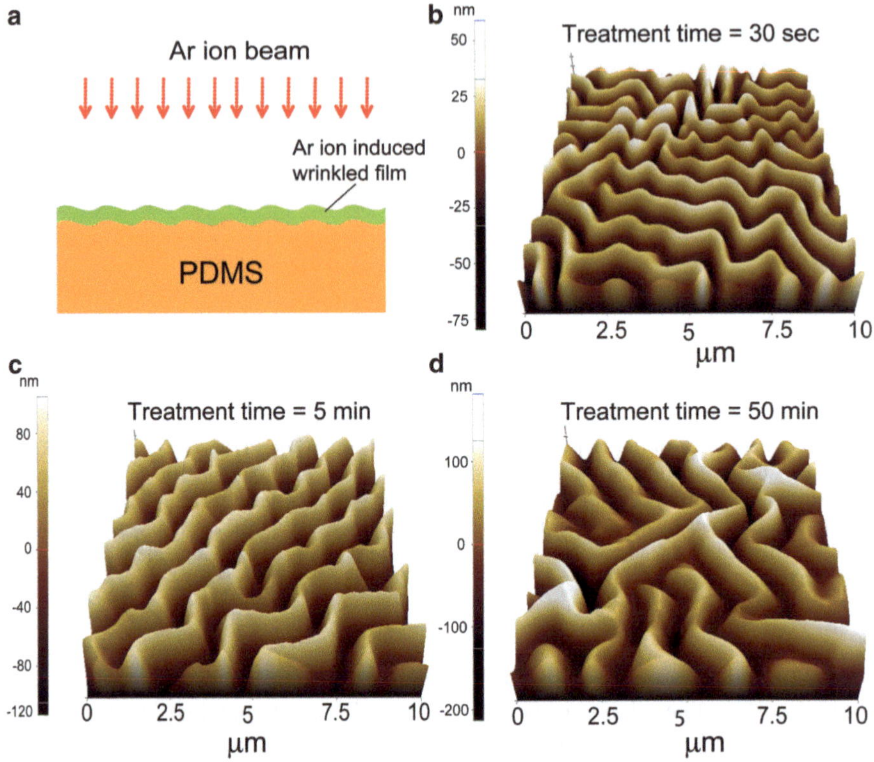

Fig. 4.13 Ar ion treatment of PDMS. (**a**) Schematic of the experiment. (**b–d**) AFM images of the wrinkles formed by the ion beam irradiation with different treatment times on the surface of PDMS. Ahmed et al. [1], reproduced by permission of The Royal Society of Chemistry. The article in which this figure was originally published is located at the following link: http://pubs.rsc.org/en/content/articlelanding/2010/sm/c0sm00386g

the pre-patterning step was varied between 10 s and 50 min. The wavelength of wrinkled patterns was controlled with the increase of pre Ar ion beam irradiation, while the amplitude was increased with deposition time of amorphous carbon layer. It was reported that for 30-min carbon deposition on PDMS with 30-s pre-patterning by Ar ion beam irradiation, the carbon layer thickness is approximately 500–600 nm, leading to an amplitude/wavelength ratio as large as 1 [1].

4.5 Summary

A brief overview of the recent surface modification technique for polymers using ion beam irradiation was provided by highlighting some of the key advantages and limitations of the developed technique to create wrinkle patterns. Firstly, methods

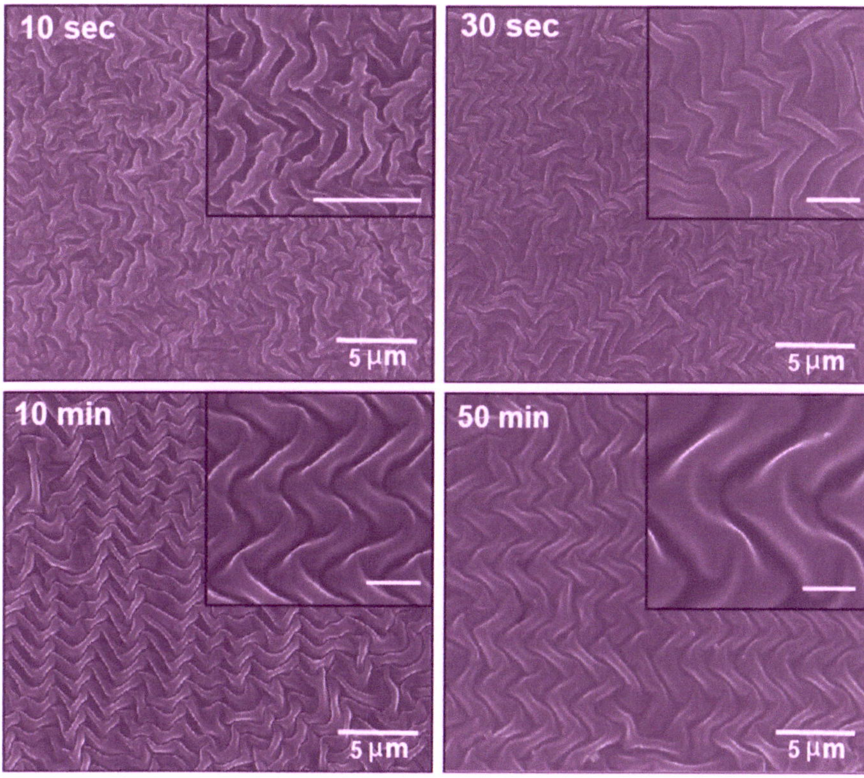

Fig. 4.14 (**a**) SEM images of the PDMS surface, pre-patterned by Ar ion with different treatment times (10 s to 50 min) and then subjected to 10-min carbon film deposition at 75° incident angles. The scale bars in the insets are 1 μm. Ahmed et al. [1], reproduced by permission of The Royal Society of Chemistry. The article in which this figure was originally published is located at the following link: http://pubs.rsc.org/en/content/articlelanding/2010/sm/c0sm00386g

to create patterns of wrinkled stiff skin on a polymeric substrate upon exposure to FIB have been demonstrated and characterized. FIB irradiation alters the chemical composition of the polymer close to its surface and induces a thin stiff skin with a strain mismatch. Moreover, some of the key experimental factors involved in patterning the polymeric surfaces by FIB were examined. Specifically, we explored the role of ion energy, pixel dwell time, and fluence on the morphology of the created wrinkles. The capabilities of the technique are also extended by adopting the maskless patterning method of the FIB system, which allows creation of self-assembled wrinkling patterns within desired surfaces of polymers.

In later part, a fabrication method of wrinkles with high aspect ratio of amplitude over wavelength was presented using a GLAD. One of the limiting factors in the usage of wrinkle patterns has been considered with the low amplitude/wavelength ratio that can be achieved using the current surface engineering techniques. This is an effective method that allows creating wrinkles with an amplitude/wavelength aspect ratio as large as 2.5 on a soft polymer.

These techniques provide robust methods for surface engineering of polymers and have direct implication in an array of multidisciplinary fields from medicine to engineering. Examples are designing biointerfaces for tissue engineering and regenerative medicine, microfluidics, biosensors, and optics.

Acknowledgements This work was supported in part by funding from KIST and in part by the US Air Force Office of Scientific Research under AFOSR YIP grant award, #FA 9550-10-1-0145, under the technical supervision of Dr. Joycelyn Harrison.

References

1. Ahmed SF, Rho GH, Lee K-R, Vaziri A, Moon M-W (2010) High aspect ratio wrinkles on a soft polymer. Soft Matter 6:5709–5714
2. Ahmed SF, Yi JW, Moon M-W, Jang Y-J, Park BH, Lee S-H, Lee K-R (2009) The morphology and mechanical properties of polycarbonate/acrylonitrile butadiene styrene modified by Ar ion beam irradiation. Plasma Proc Polym 6:860–865
3. Bowden N, Brittain S, Evans AG, Hutchinson JW, Whitesides GW (1998) Spontaneous formation of ordered structures in thin films of metals supported on an elastomeric polymer. Nature 393:146–149
4. Bradley RM, Harper JM (1988) Theory of ripple topography induced by ion bombardment. J Vac Sci Technol A 6:2390–2395
5. Briggs D, Wootton AB (1982) Analysis of polymer surfaces by SIMS 1. An investigation of practical problems. Surf Interface Anal 4:109–115
6. Chan EP, Smith EJ, Hayward RC, Crosby AJ (2008) Surface wrinkles for smart adhesion. Adv Mater 20:711–716
7. Chen X, Hutchinson JW (2004) Herringbone buckling patterns of compressed thin films on compliant substrates. ASME J Appl Mech 71:597–603
8. Chung S, Lee JH, Moon M-W, Han J, Kamm RD (2008) Non-lithographic wrinkle nanochannels for protein preconcentration. Adv Mater 20:3011–3016
9. Cerda E, Mahadevan L (2003) Geometry and physics of wrinkling. Phys Rev Lett 90:074302
10. Efimenko K, Rackaitis M, Manias E, Vaziri A, Mahadevan L, Genzer J (2005) Nested self-similar wrinkling patterns in skins. Nat Mater 4:293–297
11. El-Ali J, Sorger PK, Jensen KF (2006) Cells on chips. Nature 442:403–411
12. Erlebacher J, Aziz MJ, Chason E, Sinclair M, Floro J (1999) Spontaneous pattern formation on ion bombarded Si(001). Phys Rev Lett 82:2330–2333
13. Genzer J, Groenewold J (2006) Soft matter with hard skin: From skin wrinkles to templating and material characterization. Soft Matter 2:310–323
14. Hopman WCL, Ay F, Hu W, Gadgil VJ, Kuipers L, Pollnau M, de Ridder RRM (2007) Focused ion beam scan routine, dwell time and dose optimizations for submicrometre period planar photonic crystal components and stamps in silicon. Nanotechnology 18:195305
15. Hutchinson JW, Suo Z (1992) Mixed mode cracking in layered materials. Adv Appl Mech 29:63–191
16. Jiang X, Takayama S, Qian X, Ostuni E, Wu H, Bowden N, LeDuc P, Ingber DE, Whitesides GM (2002) Controlling mammalian cell spreading and cytoskeletal arrangement with conveniently fabricated continuous wavy features on poly (dimethylsiloxane). Langmuir 18:3273–3280

17. Kim YR, Chen P, Aziz MJ, Branton D, Vlassak JJ (2006) Focused ion beam induced deflections of freestanding thin films. J Appl Phys 100:104322
18. Krause KM, Brett MJ (2008) Spatially graded nanostructured chiral films as tunable circular polarizers. Adv Funct Mater 18:3111–3118
19. Moon M-W, Vaziri A (2009) Surface modification of polymers using a multi-step plasma treatment. Scr Mater 60:44–47
20. Moon M-W, Jensen HM, Oh KH, Hutchinson JW, Evans AG (2002) The characterization of telephone cord buckling of compressed thin films on substrates. J Mech Phys Solids 50 (11):2355–2377
21. Moon M-W, Lee K-R, Oh KH, Hutchinson JW (2004) Buckle delamination on patterned substrates. Acta Mater 52(10):3151–3159
22. Moon M-W, Lee SH, Sun JY, Oh KH, Vaziri A, Hutchinson JW (2007) Controlled formation of nanoscale wrinkling patterns on polymers using focused ion beam. Scr Mater 57 (8):747–750
23. Moon M-W, Lee SH, Sun JY, Oh KH, Vaziri A, Hutchinson JW (2007a) Wrinkled hard skins on polymers created by focused ion beam. Proc Natl Acad Sci U S A 104:1130–1133
24. Moon M-W, Han JH, Her EK, Oh KH, Lee K-R, Vaziri A, Hutchinson JW (2009) Nanoscale ripples on polymers created by a focused ion beam. Nanotechnology 20:115301
25. Moon M-W, Her EK, Oh KH, Lee KR, Vaziri A (2008) Sculpting on polymers using focused ion beam. Surf Coat Technol 202:5319–5324
26. Rahmawan Y, Moon M-W, Lee K-R, Kim K-S, Suh K-Y (2010) Wrinkled, dual-scale structures of diamond-like carbon (DLC) for superhydrophobicity. Langmuir 26:484–491
27. Satriano C, Conte E, Marletta G (2001) Surface chemical structure and cell adhesion onto ion beam modified polysiloxane. Langmuir 17:2243–2250
28. Steele JJ, Brett MJ (2007) Nanostructure engineering in porous columnar thin films: recent advances. J Mater Sci Mater Electron 18:367–379
29. Summers MA, Brett MJ (2008) Optimization of periodic column growth in glancing angle deposition for photonic crystal fabrication. Nanotechnology 19:415203
30. Yoo PJ, Suh KY, Park SY, Lee HH (2002) Physical self-assembly of microstructures by anisotropic buckling. Adv Mater 14:1383–1387

Chapter 5
A Kinetics Approach to Surface Wrinkling of Elastic Thin Films

Rui Huang

Abstract Complex wrinkle patterns have been observed in various thin film systems, typically with integrated hard and soft materials for various applications as well as in nature. The underlying mechanism of wrinkling has been generally understood as a stress-driven instability. On an elastic substrate, equilibrium and energetics set the critical condition and select the wrinkle wavelength and amplitude. On a viscous substrate, wrinkles grow over time and kinetics select the fastest growing mode. Moreover, on a viscoelastic substrate, both energetics and kinetics play important roles in determining the critical condition, the growth rate, and wrinkle patterns. The dynamics of wrinkling, while analogous to other phase ordering phenomena, is rich and distinct under the effects of stress and film–substrate interactions. In this chapter, a kinetics approach is presented for wrinkling of isotropic and anisotropic elastic films on viscoelastic substrates. Analytic solutions are obtained by a linear perturbation analysis and a nonlinear energy minimization method, which predict the kinetics of wrinkle growth at the early stage and the equilibrium states at the long-time limit, respectively. In between, a power-law coarsening of the wrinkle wavelength is predicted by a scaling analysis. Furthermore, the kinetics approach enables numerical simulations that demonstrate emergence and transition of diverse wrinkle patterns (ordered and disordered) under various conditions.

R. Huang (✉)
Department of Aerospace Engineering and Engineering Mechanics, University of Texas at Austin, Austin, TX 78712, USA
e-mail: ruihuang@mail.utexas.edu

X. Chen (ed.), *Mechanical Self-Assembly: Science and Applications*,
DOI 10.1007/978-1-4614-4562-3_5, © Springer Science+Business Media New York 2013

5.1 Introduction

Theoretical studies of surface wrinkling may be traced back to 1940s when wrinkling of face struts was analyzed as a form of local elastic instability in structural sandwich panels [1–3]; an account of the historical development was well documented by Allen [4]. A series of works by Biot extended the wrinkling theory to viscoelastic layers [5] and rubberlike nonlinear elastic media under finite strain [6]. The early studies of wrinkling focused on the critical conditions for the onset of instability as a failure mechanism for layered structures. Recent advances in micro/nanoscale fabrication and measurements have renewed the interest in mechanics of wrinkling. In particular, wrinkling has also been exploited as an enabling mechanism for a variety of applications, such as stretchable electronics [7–9], micro/nanofabrication [10–12], optical phase grating [13], smart adhesion [14], and metrology aid for measuring mechanical properties of thin films [15–17].

Wrinkling often occurs when a stiff surface layer on a drastically more compliant substrate is subject to compression in a direction parallel to the surface. Many approaches may induce compression and wrinkling. Mechanically, the surface layer can be put under compression by directly compressing the substrate or by bending the substrate. Interestingly, stretching the substrate could also induce compression in the surface layer [18] due to Poisson's effect. Typically the substrate (e.g., elastomer) has a larger Poisson's ratio than the surface layer (e.g., metal), and the mismatch induces compression of the surface layer in the direction perpendicular to the direction of stretch. More effectively, by attaching the surface layer to a pre-stretched substrate, the surface layer is compressed upon releasing the pre-stretch and the resulting wrinkles are stable without the need to apply external forces. Moreover, many nonmechanical approaches have also been reported to cause wrinkling, which often relies on a particular physical mechanism to induce an eigenstrain in the surface layer. A common approach of this kind is thermally induced wrinkling [10, 17, 19, 20], with the eigenstrain resulting from differential thermal expansion among the layers. A similar mechanism in biological systems is differential tissue growth in skin layers [21, 22] and human brain [23]. Other mechanisms include absorption-induced swelling [24] and phase transition [25]. In general, one may assume a residual stress in the surface layer as the driving force for wrinkling, regardless of the physical origin.

Notably, surface wrinkling may occur without any surface layer [26–28]. As predicted by Biot [6], the surface of a homogeneous block of rubberlike material becomes unstable under compression beyond a moderately large strain. Similar surface instability has been predicted for swollen hydrogels [29]. This type of surface instability, however, requires a nonlinear constitutive behavior of the material and is not addressed here.

In this chapter, we focus on a model system as shown in Fig. 5.1: an elastic film of thickness h lying on a viscoelastic layer of thickness H, which in turn lies on a rigid foundation. At the reference state (Fig. 5.1a), both layers are flat and the elastic film is subjected to a uniformly distributed residual stress, while the

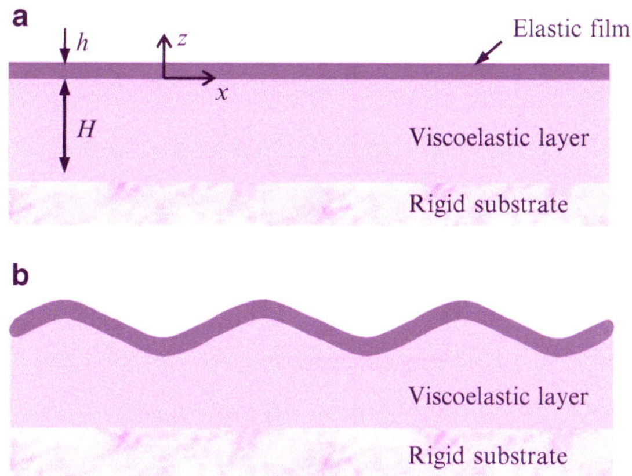

Fig. 5.1 Schematic of an elastic film on a viscoelastic substrate: (**a**) the reference state, and (**b**) a wrinkled state

viscoelastic layer is stress free. A Cartesian coordinate system is set up such that the $x-y$ plane coincides with the interface between the two layers. Figure 5.1b sketches a wrinkled state, where the elastic film undergoes a buckling deformation while the viscoelastic layer deforms concomitantly. The interface is assumed to be perfectly bonded at the wrinkled state.

Consideration of a viscoelastic layer underneath an elastic film leads to a kinetics approach to wrinkling [30–34]. Alternatively, an energy-based statics approach has been widely adopted for post-instability analysis to determine the equilibrium states of wrinkling for an elastic film on an elastic substrate [35–40]. Compared to the static equilibrium approach, which typically presumes specific wrinkle patterns, the kinetics approach offers a physical pathway to a variety of ordered and disordered wrinkle patterns without a priori assumptions. A similar kinetics approach has also been developed for wrinkling of an elastic film on a viscous substrate [41–44].

5.2 Deformation and Equilibrium of an Elastic Film

Consider a cubic crystal film with the surface normal in the [001]-direction. For convenience, the x- and y-axes are set to align with the [100]- and [010]-directions of the crystal at the reference state. The residual stress in the film generally has three in-plane components, σ_{xx}^{R}, σ_{yy}^{R}, and σ_{xy}^{R}, as illustrated in Fig. 5.2a. The stress state can also be represented by two principal stresses (σ_1 and σ_2) and the corresponding principal angle (θ_p), as illustrated in Fig. 5.2b. As a necessary condition for the film to wrinkle, at least one of the two principal stresses must be negative (compressive). The resulting wrinkle pattern depends on the ratio between the two principal stresses for an isotropic film. For an anisotropic crystal film, the wrinkle pattern also depends on the principal direction of the residual stress [33].

Let u_x and u_y be the in-plane displacements and w the out-of-plane deflection of the film. Upon wrinkling, the membrane strain components in the film are

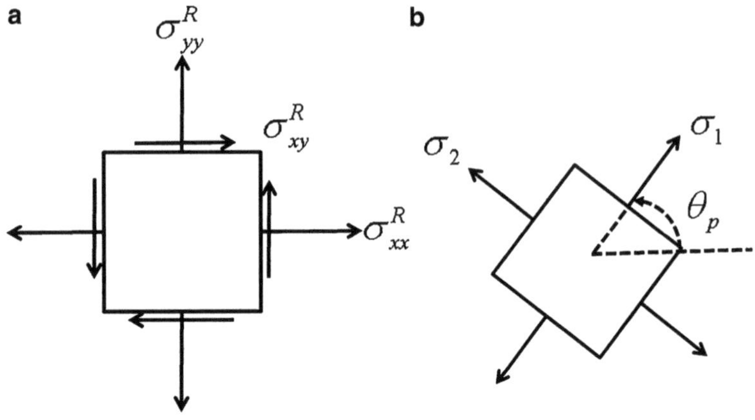

Fig. 5.2 Residual stress in a film represented by stress elements: (**a**) in the *x*–*y* coordinates; (**b**) in the principal directions

$$\varepsilon_{xx} = \frac{\partial u_x}{\partial x} + \frac{1}{2}\left(\frac{\partial w}{\partial x}\right)^2, \tag{5.1}$$

$$\varepsilon_{yy} = \frac{\partial u_y}{\partial y} + \frac{1}{2}\left(\frac{\partial w}{\partial y}\right)^2, \tag{5.2}$$

$$\varepsilon_{xy} = \frac{1}{2}\left(\frac{\partial u_x}{\partial y} + \frac{\partial u_y}{\partial x}\right) + \frac{1}{2}\frac{\partial w}{\partial x}\frac{\partial w}{\partial y}. \tag{5.3}$$

A nonlinear term is included above in each strain component to account for geometrical nonlinearity due to moderately large deflection of the elastic film [45, 46]. It shall be noted that the displacements are measured relative to the reference state with the residual stress $\sigma^{\mathrm{R}}_{\alpha\beta}$, and thus the strains in (5.1)–(5.3) describe an incremental deformation from the stressed reference state.

The membrane forces in the film, including the initial residual stresses, are

$$N_{\alpha\beta} = \left(\sigma^{\mathrm{R}}_{\alpha\beta} + \sigma_{\alpha\beta}\right)h, \tag{5.4}$$

where the Greek subscripts (α, β) represent the in-plane coordinates x and y, and by linear elasticity

$$\begin{bmatrix} \sigma_{xx} \\ \sigma_{yy} \\ \sigma_{xy} \end{bmatrix} = \begin{bmatrix} \tilde{C}_{11} & \tilde{C}_{12} & 0 \\ \tilde{C}_{12} & \tilde{C}_{22} & 0 \\ 0 & 0 & 2C_{66} \end{bmatrix} \begin{bmatrix} \varepsilon_{xx} \\ \varepsilon_{yy} \\ \varepsilon_{xy} \end{bmatrix}. \tag{5.5}$$

For a cubic crystal film, the elastic moduli are $\tilde{C}_{11} = \tilde{C}_{22} = C_{11} - C_{12}^2/C_{11}$ and $\tilde{C}_{12} = C_{12} - C_{12}^2/C_{11}$. For an isotropic elastic plate, the reduced elastic constants can be expressed in terms of Young's modulus (E) and Poisson's ratio (ν), namely, $\tilde{C}_{11} = \tilde{C}_{22} = E/(1-\nu^2)$, $\tilde{C}_{12} = \nu E/(1-\nu^2)$, and $C_{66} = E/[2(1+\nu)]$.

The moment intensities are

$$\begin{bmatrix} M_{xx} \\ M_{yy} \\ M_{xy} \end{bmatrix} = \frac{h^3}{12} \begin{bmatrix} \tilde{C}_{11} & \tilde{C}_{12} & 0 \\ \tilde{C}_{12} & \tilde{C}_{22} & 0 \\ 0 & 0 & 2C_{66} \end{bmatrix} \begin{bmatrix} \kappa_{xx} \\ \kappa_{yy} \\ \kappa_{xy} \end{bmatrix}, \tag{5.6}$$

where $\kappa_{\alpha\beta} = \frac{\partial^2 w}{\partial x_\alpha \partial x_\beta}$ is the curvature tensor.

For the film to be in equilibrium at the wrinkled state, interfacial tractions are generally required in both the tangential and normal directions. First, in-plane force equilibrium requires that

$$\frac{\partial N_{xx}}{\partial x} + \frac{\partial N_{xy}}{\partial y} = T_x, \tag{5.7}$$

$$\frac{\partial N_{xy}}{\partial x} + \frac{\partial N_{yy}}{\partial y} = T_y, \tag{5.8}$$

where T_x and T_y are the shear tractions acting at the interface between the film and the substrate. In the normal direction, the force equilibrium requires that

$$\frac{\partial Q_x}{\partial x} + \frac{\partial Q_y}{\partial y} + \left[\frac{\partial}{\partial x}\left(N_{xx}\frac{\partial w}{\partial x} + N_{xy}\frac{\partial w}{\partial y}\right) + \frac{\partial}{\partial y}\left(N_{yy}\frac{\partial w}{\partial y} + N_{xy}\frac{\partial w}{\partial x}\right)\right] + p = 0, \tag{5.9}$$

where p is the pressure (negative normal traction) at the interface, Q_x and Q_y are the transverse shearing forces in the film. In addition, by the moment equilibrium, we have

$$\frac{\partial M_{xx}}{\partial x} + \frac{\partial M_{xy}}{\partial y} + Q_x = 0, \tag{5.10}$$

$$\frac{\partial M_{xy}}{\partial x} + \frac{\partial M_{yy}}{\partial y} + Q_y = 0. \tag{5.11}$$

With (5.10) and (5.11), the shear forces in (5.9) can be eliminated, namely,

$$\frac{\partial^2 M_{xx}}{\partial x^2} + 2\frac{\partial^2 M_{xy}}{\partial x \partial y} + \frac{\partial^2 M_{yy}}{\partial y^2} - \left[\frac{\partial}{\partial x}\left(N_{xx}\frac{\partial w}{\partial x} + N_{xy}\frac{\partial w}{\partial y}\right) + \frac{\partial}{\partial y}\left(N_{yy}\frac{\partial w}{\partial y} + N_{xy}\frac{\partial w}{\partial x}\right)\right] = p. \tag{5.12}$$

Further by substituting (5.6) into (5.12), we obtain that

$$\frac{h^3}{12}\left[\tilde{C}_{11}\frac{\partial^4 w}{\partial x^4}+2(\tilde{C}_{12}+2C_{66})\frac{\partial^4 w}{\partial x^2\partial y^2}+\tilde{C}_{22}\frac{\partial^4 w}{\partial y^4}\right] \\ -\left[\frac{\partial}{\partial x}\left(N_{xx}\frac{\partial w}{\partial x}+N_{xy}\frac{\partial w}{\partial y}\right)+\frac{\partial}{\partial y}\left(N_{yy}\frac{\partial w}{\partial y}+N_{xy}\frac{\partial w}{\partial x}\right)\right]=p. \tag{5.13}$$

For an isotopic elastic film, (5.13) reduces to

$$\frac{Eh^3}{12(1-\nu^2)}\left(\frac{\partial^4 w}{\partial x^4}+2\frac{\partial^4 w}{\partial x^2\partial y^2}+\frac{\partial^4 w}{\partial y^4}\right) \\ -\left[\frac{\partial}{\partial x}\left(N_{xx}\frac{\partial w}{\partial x}+N_{xy}\frac{\partial w}{\partial y}\right)+\frac{\partial}{\partial y}\left(N_{yy}\frac{\partial w}{\partial y}+N_{xy}\frac{\partial w}{\partial x}\right)\right]=p, \tag{5.14}$$

which is well known as the von-Karman plate equation [45, 46].

Therefore, the deformation of the elastic film is described by the displacements (u_x, u_y, w), the membrane strain ($\varepsilon_{\alpha\beta}$), and the curvature ($\kappa_{\alpha\beta}$). The constitutive relations in (5.4–5.6) give the membrane forces ($N_{\alpha\beta}$) and moments ($M_{\alpha\beta}$), assuming linear elasticity. The tractions (T_x, T_y, p) at the interface are then obtained from the equilibrium equations in (5.7), (5.8), and (5.12).

5.3 Kinetics: Deformation of a Viscoelastic Layer

Subject to the interfacial tractions, the viscoelastic layer undergoes a time-dependent deformation. By the linear theory of viscoelasticity [47], the stress–strain relationship for the viscoelastic material takes an integral form:

$$\sigma_{ij}(t)=2\int_{-\infty}^{t}\mu(t-\tau)\frac{\partial\varepsilon_{ij}(\tau)}{\partial\tau}\mathrm{d}\tau+\delta_{ij}\int_{-\infty}^{t}\lambda(t-\tau)\frac{\partial\varepsilon_{kk}(\tau)}{\partial\tau}\mathrm{d}\tau, \tag{5.15}$$

where $\mu(t)$ and $\lambda(t)$ are the viscoelastic relaxation moduli and δ_{ij} is the Kronecker delta.

The Laplace transform of (5.15) is

$$\tilde{\sigma}_{ij}(s)=2s\tilde{\mu}(s)\tilde{\varepsilon}_{ij}(s)+s\tilde{\lambda}_{ij}(s)\tilde{\varepsilon}_{kk}(s)\delta_{ij}, \tag{5.16}$$

where a tilt over a variable designates its Laplace transform with respect to time and s is the transform variable. The Laplace transformed stress–strain relation is identical to that of linear elasticity with the elastic moduli $s\tilde{\mu}(s)$ and $s\tilde{\lambda}(s)$. As a result, a viscoelasticity problem often can be solved by the method of Laplace transform based on the solution of a corresponding elasticity problem, referred to as the correspondence principle [47].

Consider a viscoelastic layer ($0 \geq z \geq -H$), which is stress free initially ($t = 0$) and subjected to normal and shear tractions at the upper surface for $t>0$, namely,

$$\sigma_{zz} = S_z(x, y, t) \text{ and } \sigma_{z\alpha} = S_\alpha(x, y, t) \text{ at } z = 0. \tag{5.17}$$

The lower surface of the layer is bonded to a rigid substrate, where the displacement is zero:

$$u_\alpha = u_z = 0 \text{ at } z = -H. \tag{5.18}$$

Assume a plane-strain condition ($S_y = u_y = 0$) with periodic surface tractions in the form

$$S_x = A(t) \sin kx, \tag{5.19}$$

$$S_z = B(t) \cos kx, \tag{5.20}$$

where k is the wave number and $A(t)$ and $B(t)$ are time-dependent amplitudes.

By solving the corresponding elasticity problem [30], the Laplace transform of the displacements at the surface of the viscoelastic layer is obtained as

$$\tilde{u}_x^s(x, s) = \frac{1}{2ks\tilde{\mu}(s)} \left[\gamma_{11}(s\tilde{v}, kH)\tilde{A}(s) + \gamma_{12}(s\tilde{v}, kH)\tilde{B}(s)\right] \sin(kx), \tag{5.21}$$

$$\tilde{u}_z^s(x, s) = \frac{1}{2ks\tilde{\mu}(s)} \left[\gamma_{21}(s\tilde{v}, kH)\tilde{A}(s) + \gamma_{22}(s\tilde{v}, kH)\tilde{B}(s)\right] \cos(kx), \tag{5.22}$$

where

$$\gamma_{11} = \frac{1+\kappa}{4} \frac{\kappa \sinh(2kH) + 2kH}{\kappa \cosh^2(kH) + (kH)^2 + ((1-\kappa)/2)^2}, \tag{5.23}$$

$$\gamma_{22} = \frac{1+\kappa}{4} \frac{\kappa \sinh(2kH) - 2kH}{\kappa \cosh^2(kH) + (kH)^2 + ((1-\kappa)/2)^2}, \tag{5.24}$$

$$\gamma_{12} = \gamma_{21} = -\frac{(\kappa(1-\kappa))/2\sinh^2(kH) + (kH)^2}{\kappa \cosh^2(kH) + (kH)^2 + ((1-\kappa)/2)^2}, \tag{5.25}$$

and $\kappa = 3 - 4s\tilde{v}(s)$ with $\tilde{v}(s) = \tilde{\lambda}(s)/(2s(\tilde{\mu}(s) + \tilde{\lambda}(s)))$ as Laplace transform of Poisson's ratio.

The solution in (5.21) and (5.22) indicates that, in general, the surface of the viscoelastic layer undergoes both out-of-plane and in-plane displacements. For a special case, when the layer is infinitely thick ($kH \to \infty$) and incompressible ($v = 0.5$), we have $\gamma_{12} = \gamma_{21} = 0$, and thus the two equations are decoupled:

$$\tilde{u}_x^s(x,s) = \frac{1}{4ks\tilde{\mu}(s)}\tilde{S}_x(x,s), \tag{5.26}$$

$$\tilde{u}_z^s(x,s) = \frac{1}{4ks\tilde{\mu}(s)}\tilde{S}_z(x,s). \tag{5.27}$$

Alternatively, if the viscoelastic layer is relatively thin ($kH \to 0$), by keeping only the leading terms of kH in an asymptotic analysis [31, 34], (5.21) and (5.22) become

$$\tilde{u}_x^s(x,s) = \frac{1}{2ks\tilde{\mu}(s)}\left[2kH\tilde{A}(s) + \frac{1-4v}{2(1-v)}(kH)^2\tilde{B}(s)\right]\sin(kx), \tag{5.28}$$

$$\tilde{u}_z^s(x,s) = \frac{1}{2ks\tilde{\mu}(s)}\left[\frac{1-2v}{1-v}(kH)\tilde{B}(s) + \frac{1-4v}{2(1-v)}(kH)^2\tilde{A}(s)\right]\cos(kx). \tag{5.29}$$

Here, the Poisson's ratio has been assumed to be a constant independent of time and $v < 0.5 - (1/6)(kH)^2$ [34]. On the other hand, when the Poisson's ratio approaches 0.5, the viscoelastic layer is nearly incompressible and (5.29) is replaced by

$$\tilde{u}_z^s(x,s) = \frac{1}{2ks\tilde{\mu}(s)}\left[\frac{2}{3}(kH)^3\tilde{B}(s) - (kH)^2\tilde{A}(s)\right]\cos(kx), \tag{5.30}$$

where the first term in the bracket scales with $(kH)^3$ instead of (kH) in (5.29). This leads to different kinetics of wrinkling for compressible and incompressible viscoelastic layers [32].

Figure 5.3 depicts the relaxation shear modulus, $\mu(t)$, as a function of time for a typical cross-linked polymer, with the glassy modulus $\mu(0) = \mu_G$ at the short-time limit and the rubbery modulus $\mu(\infty) = \mu_R$ at the long-time limit. The modulus typically varies by several orders of magnitude: $\mu_G \sim 10^9$ Pa and $\mu_R \sim 10^5$ Pa. In general, the Poisson's ratio is also a function of time, but the time dependence is much weaker. Experimentally measured relaxation modulus $\mu(t)$ is often interpreted in terms of a mechanical model consisting of an array of spring-dashpot analogs in parallel (Fig. 5.4). To ensure a rubbery elastic limit at the long-time limit, one of the parallel branches must be a spring of modulus μ_R, with no dashpot. Each of the other parallel branches comprises a spring of modulus μ_i and a dashpot of viscosity η_i. With such a model, the relaxation modulus is

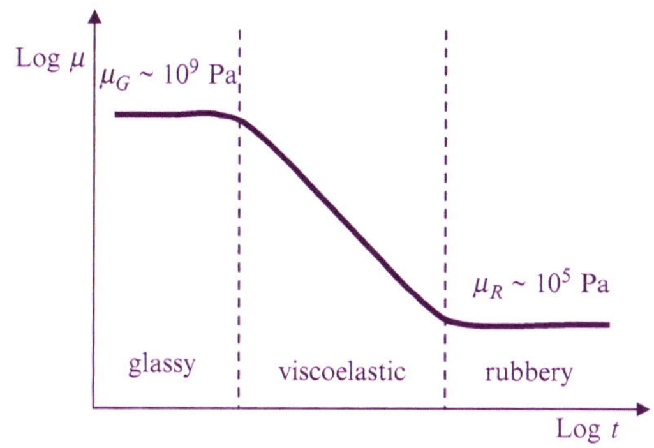

Fig. 5.3 A schematic of the relaxation modulus as a function of time for a viscoelastic material

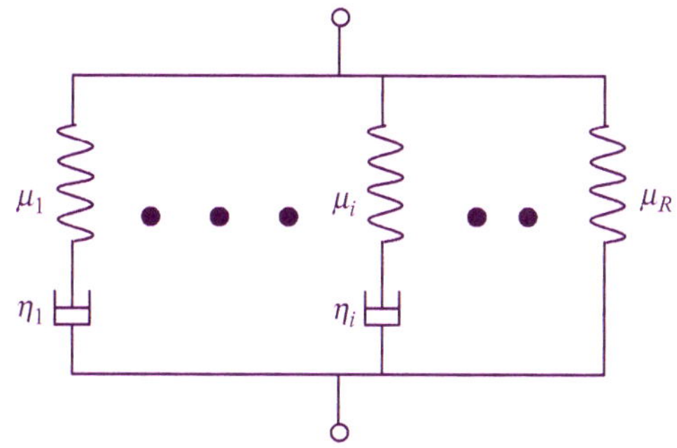

Fig. 5.4 A mechanical analog model of a viscoelastic material

$$\mu(t) = \mu_{\mathrm{R}} + \sum_i \mu_i \exp(-p_i t), \tag{5.31}$$

where $p_i = \mu_i/\eta_i$ is the relaxation parameter of one branch. The Laplace transform of the relaxation modulus is

$$\tilde{\mu}(s) = \frac{\mu_{\mathrm{R}}}{s} + \sum_i \frac{\mu_i}{s + p_i}. \tag{5.32}$$

For simplicity, consider the Kelvin model of linear viscoelasticity, modeled by a simple mechanical analog consisting of an elastic spring and a viscous dashpot in parallel, for which the Laplace transform of the relaxation modulus is

$$\tilde{\mu}(s) = \frac{\mu_{\mathrm{R}}}{s} + \eta. \tag{5.33}$$

After substituting (5.33) into (5.28) and (5.29), inverse Laplace transform leads to

$$\frac{\partial u_x^s}{\partial t} = \frac{H}{\eta} S_x + \frac{1-4\nu}{4(1-\nu)} \frac{H^2}{\eta} \frac{\partial S_z}{\partial x} - \frac{\mu_{\mathrm{R}}}{\eta} u_x^s, \tag{5.34}$$

$$\frac{\partial u_z^s}{\partial t} = \frac{1-2\nu}{2(1-\nu)}\frac{H}{\eta}S_z - \frac{1-4\nu}{4(1-\nu)}\frac{H^2}{\eta}\frac{\partial S_x}{\partial x} - \frac{\mu_R}{\eta}u_z^s. \tag{5.35}$$

By dropping the H^2 terms under the thin-layer approximation ($kH \ll 1$), (5.34) and (5.35) become

$$\frac{\partial u_x^s}{\partial t} = \frac{H}{\eta}S_x - \frac{\mu_R}{\eta}u_x^s, \tag{5.36}$$

$$\frac{\partial u_z^s}{\partial t} = \frac{1-2\nu}{2(1-\nu)}\frac{H}{\eta}S_z - \frac{\mu_R}{\eta}u_z^s. \tag{5.37}$$

Equation (5.36) is equivalent to a shear-lag model [48–50], which assumes a uniform shear stress across the thin layer. Equation (5.37) is similar to the Winkler model for an elastic foundation [4], but includes a time-dependent term accounting for the viscous effect. The two equations are uncoupled under the thin-layer approximation.

In the above development, plane-strain deformation and periodic surface tractions have been assumed. The restriction of periodic tractions has been relaxed by absorbing the wave number through a differentiation with respect to x in the course of the inverse Laplace transform. The resulting equations are independent of wave number and can thus be used for arbitrary tractions. At the end, the in-plane and out-of-plane responses are decoupled by the thin-layer approximation. Therefore, the restriction of plane-strain deformation can also be relaxed by generalizing the in-plane response, (5.36), for both x- and y-directions, namely,

$$\frac{\partial u_\alpha^s}{\partial t} = \frac{H}{\eta}S_\alpha - \frac{\mu_R}{\eta}u_\alpha^s, \tag{5.38}$$

for $\alpha = x$ and y. Equations (5.37) and (5.38) then represent an approximate solution for the three-dimensional response of a thin viscoelastic layer subjected to the boundary conditions in (5.17) and (5.18). Following similar steps, other forms of viscoelastic solutions may be developed for incompressible and thick layers.

5.4 Evolution Equations of Wrinkling

By assuming the interface between the elastic film and the viscoelastic layer to remain bonded during wrinkling, the displacements and tractions are continuous across the interface, namely,

$$S_z = -p, \quad S_\alpha = T_\alpha, \tag{5.39}$$

$$u_z^s = w, \quad u_\alpha^s = u_\alpha. \tag{5.40}$$

The time-dependent response of the viscoelastic layer can then be written in terms of the film displacements as

$$\frac{\partial w}{\partial t} = \frac{1-2\nu}{2(1-\nu)} \frac{H}{\eta}(-p) - \frac{\mu_R}{\eta} w, \tag{5.41}$$

$$\frac{\partial u_\alpha}{\partial t} = \frac{H}{\eta} T_\alpha - \frac{\mu_R}{\eta} u_\alpha, \tag{5.42}$$

where p and T_α are related to the displacements (w and u_α) by (5.7), (5.8), and (5.12). Together with the constitutive relations of the film in (5.4)–(5.6), we have a complete set of evolution equations for wrinkling.

Here, it is assumed that the evolution process is sufficiently slow such that inertia effects are negligible and the elastic film remains in equilibrium during evolution. Despite the limitations in the simple viscoelastic property and the thin-layer approximation, the present model has been shown to be able to capture essential features of viscoelastic wrinkle evolution, such as the kinetics of wrinkle growth at the early stage and the equilibrium states at the long-time limit. Extension to more general cases is possible. Notably, Biot [5] developed a general instability theory for a viscoelastic layer on a semi-infinite viscoelastic substrate or embedded in an infinite viscoelastic medium, which may be extended to three-dimensional post-instability analysis. Alternatively, a Fourier transform method similar to that by Huang et al. [37] may be employed for an elastic film on an infinitely thick viscoelastic substrate. By combining Reynolds' lubrication theory with the nonlinear plate equations, Huang and Suo [42] developed a set of evolution equations for wrinkling of an elastic film on an incompressible viscous layer.

5.5 Initial Growth

Start from a nearly flat film. At the early stage of wrinkling, the residual stress in the film is hardly relaxed, and the growth of wrinkles is predominantly due to out-of-plane displacement, uncoupled from the in-plane displacements. Substituting (5.14) into (5.41) for an isotropic film, we obtain that

$$\frac{\partial w}{\partial t} = -Kw_{,\alpha\alpha\beta\beta} + F\left(N_{\alpha\beta}w_{,\beta}\right)_{,\alpha} - Rw, \tag{5.43}$$

where

$$K = \frac{(1-2\nu_s)E_f h^3 H}{24(1-\nu_s)(1-\nu_f^2)\eta_s}, \quad F = \frac{1-2\nu_s}{2(1-\nu_s)} \frac{H}{\eta_s}, \quad R = \frac{\mu_R}{\eta_s}. \tag{5.44}$$

Consider an equi-biaxial residual stress such that $N_{\alpha\beta} \approx -\sigma_0 h \delta_{\alpha\beta}$ ($\sigma_0 > 0$) at the early stage. The first two terms at the right-hand side of (5.43) compete to set a length scale

$$L_1 = \sqrt{\frac{K}{F\sigma_0 h}} = \sqrt{\frac{E_f h^2}{12(1-\nu_f^2)\sigma_0}}, \tag{5.45}$$

and a corresponding time scale

$$\tau_1 = \frac{K}{(F\sigma_0 h)^2} = \frac{(1-\nu_s)E_f \eta_s h}{6(1-2\nu_s)(1-\nu_f^2)\sigma_0^2 H}. \tag{5.46}$$

Neglecting the third term at the right-hand side of (5.43) for the moment, a linear perturbation analysis [31] leads to a critical wavelength,

$$\lambda_c = 2\pi L_1, \tag{5.47}$$

and the fastest growing wavelength,

$$\lambda_m = 2\sqrt{2}\pi L_1, \tag{5.48}$$

both proportional to the length scale, which in turn is proportional to the film thickness. The film is unstable for wrinkles with wavelengths $\lambda > \lambda_c$, and the wrinkle amplitude grows exponentially with time. The growth rate, inversely proportional to the time scale τ_1, peaks at the wavelength $\lambda = \lambda_m$, for which the wrinkle amplitude is given by

$$A = A_0 \exp\left(\frac{t}{4\tau_1}\right), \tag{5.49}$$

and A_0 is a constant for the initial amplitude. Therefore, at the early stage, the fastest growing mode dominates, with both the wrinkle wavelength and the growth rate depending on the residual stress via the length and time scales.

The third term at the right-hand side of (5.43) accounts for the effect of substrate elasticity, which does not change the fastest growing wrinkle wavelength but reduces the growth rate [31]. The substrate elasticity completely suppresses the wrinkling instability when the compressive residual stress is lower than a critical value. By setting the peak growth rate to be zero, one obtains the critical stress:

$$\sigma_c = \sqrt{\frac{4KR}{F^2 h^2}} = \sqrt{\frac{2(1-\nu_s)E_f \mu_R}{3(1-2\nu_s)(1-\nu_f^2)}\frac{h}{H}}. \tag{5.50}$$

Note that the critical stress in (5.50) decreases as the thickness ratio H/h increases, but this is only true for a relatively thin viscoelastic layer. For a thick substrate layer, the critical stress approaches a constant independent of the thickness ratio [30, 37].

For an anisotropic elastic film, a linear perturbation analysis predicts anisotropic wrinkle patterns at the early stage [33]. Consider a periodic wrinkle pattern, with a wave number k and an angle of the wave vector θ, namely,

$$w(x, y, t) = A(t)\cos[k(x\cos\theta + y\sin\theta)]. \tag{5.51}$$

Here, the angle θ is measured from the x-axis or the [100] axis for the cubic crystal film. The wrinkle amplitude grows exponentially with time as

$$A(t) = A_0 \exp\left(\frac{\alpha_\theta t}{\tau}\right), \tag{5.52}$$

where $\tau = \eta_s/C_{11}$ is a time scale, and

$$\alpha_\theta = \frac{(1-2\nu_s)H}{24(1-\nu_s)C_{11}h}\left[-\overline{E}_\theta(kh)^4 - 12\sigma_\theta(kh)^2\right] - \frac{\mu_R}{C_{11}}. \tag{5.53}$$

The dimensionless growth rate in (5.53) depends on the angle θ through two quantities:

$$\sigma_\theta = \sigma_1\left[1 - \left(1 - \frac{\sigma_2}{\sigma_1}\right)\sin^2(\theta - \theta_p)\right], \tag{5.54}$$

$$\overline{E}_\theta = C_{11}\left[1 - \left(\frac{C_{12}}{C_{11}}\right)^2\right]\left[1 + \frac{(\xi - 1)C_{11}}{2(C_{11}+C_{12})}\sin^2 2\theta\right]. \tag{5.55}$$

First, the ratio between the two principal stresses represents the stress anisotropy that determines the angular dependence of σ_θ. Hence an equi-biaxial stress ($\sigma_2/\sigma_1 = 1$) is isotropic. Second, $\xi = 2C_{66}/(C_{11} - C_{12})$ defines a degree of elastic anisotropy of the crystal. For an isotropic material, $\xi = 1$ and $\overline{E}_\theta$ reduces to $\overline{E}_f = E_f/(1 - \nu_f^2)$, independent of the angle.

Figure 5.5 compares the spectra of the initial wrinkle growth rate for both isotropic and anisotropic elastic films under various residual stresses [33]. Contours of the normalized growth rate α_θ are plotted in the plane spanning the x and y components of the wave vector, $k_x = k\cos\theta$ and $k_y = k\sin\theta$, both normalized by the film thickness h; only positive growth rates are plotted. In all the calculations, we set $\mu_R/C_{11} = 10^{-5}$, $H/h = 10$, and $\nu_s = 0.45$. The major principal stress $\sigma_1/C_{11} = -0.003$ is fixed, while the ratio σ_2/σ_1 is varied from 1 for equi-biaxial to 0 for uniaxial stress state. For the isotropic film, the growth

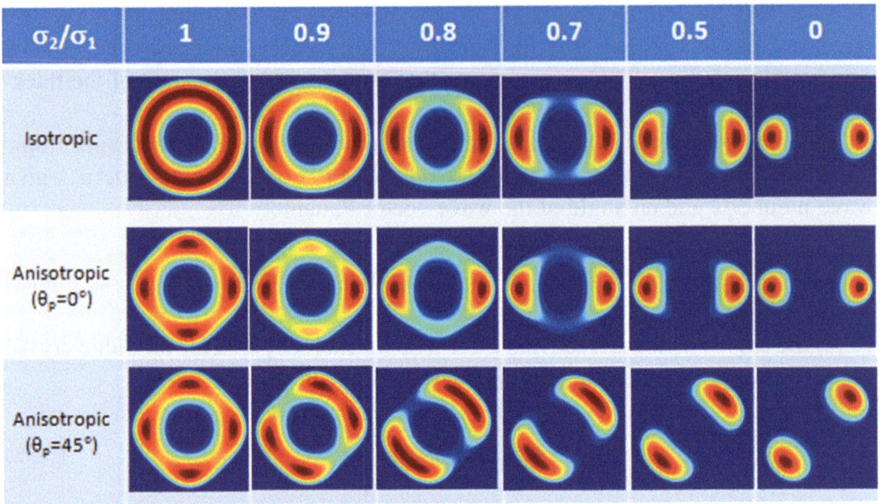

Fig. 5.5 Spectra of the wrinkle growth rate at the early stage for isotropic and anisotropic elastic films under various residual stress states. The major principal stress is $\sigma_1/C_{11} = -0.003$

spectrum is solely controlled by the stress ratio. Under an equi-biaxial residual stress $(\sigma_2/\sigma_1 = 1)$, the growth spectrum is isotropic with concentric circular contours and thus no favored directions. When the two principal stresses differ, the rotational symmetry is broken and the growth rate peaks at a particular wave vector in the direction of the major principal stress. Therefore, the kinetically dominant wrinkle pattern at the initial stage changes from nondirectional (e.g., labyrinth pattern) to unidirectional (e.g., parallel striped pattern).

For a cubic crystal film (e.g., SiGe), the growth spectrum not only depends on the ratio between the two principal stresses but also depends on the direction of the principal stress (θ_p). Even under an isotropic equi-biaxial stress, an anisotropic growth spectrum emerges, with four peaks aligned in the two orthogonal crystal directions, [100] and [010]. Thus, while the stress state is isotropic, the anisotropic elastic property of the crystal film breaks the rotational symmetry. As a result, an orthogonally oriented bidirectional pattern is predicted to dominate the initial growth of the wrinkles. By varying the ratio σ_2/σ_1 from 1 to 0, the growth spectrum changes from orthogonal to uniaxial in the direction of the major principal stress. The transition however depends on the principal angle θ_p. When $\theta_p = 0$, the growth spectrum remains orthogonal when the ratio σ_2/σ_1 slightly deviates from 1, but the peak growth rate becomes lower in one direction compared to the other. The growth spectrum becomes uniaxial as the lower peak diminishes for $\sigma_2/\sigma_1 < 0.8$. When $\theta_p = 45°$, the wave vectors corresponding to the peak growth rates rotate toward the direction of the major principal stress and eventually merge into a uniaxial pattern. At an intermediate stress ratio (e.g., $\sigma_2/\sigma_1 = 0.9$), four peaks lie on two directions of an oblique angle. Hence, the kinetically dominant wrinkle pattern at the early stage becomes obliquely oriented bidirectional (e.g., zigzag pattern).

This transition in the initial wrinkle patterns may be understood as a result of the competition between the material anisotropy and the stress anisotropy through $\overline{E}_\theta$ and σ_θ, respectively.

5.6 Equilibrium Wrinkles

For a viscoelastic substrate layer with a rubbery modulus $\mu_R>0$ at the long-time limit, wrinkling of an elastic film atop evolves toward an equilibrium state, dictated by minimization of the elastic strain energy stored in the film and the substrate. There may exist many mechanically equilibrium states, including the one at the reference state with no wrinkles at all. However, by the principle of thermodynamics these equilibrium states may be unstable, stable, or metastable. Searching for the thermodynamically equilibrium state with the minimum energy requires consideration of all possible wrinkle patterns. In practice, several simple wrinkle patterns (e.g., parallel stripes, checkerboard, and herringbone) have been considered for isotropic elastic films [36, 39]. However, experiments have commonly observed disordered wrinkle patterns (e.g., labyrinth). With the kinetics approach facilitated by the viscoelastic deformation in the substrate layer, evolution of wrinkle patterns can be simulated from a randomly generated initial perturbation, without a priori assumption of the wrinkle patterns. Since the viscoelastic deformation dissipates energy, the evolution process may be regarded as a searching algorithm for the minimum energy state, but it is not guaranteed that the global energy minimum can be reached. Nevertheless, the viscoelastic evolution represents a common physical process to form wrinkle patterns in experiments [17, 19, 51, 52], where the observed wrinkle patterns may also be trapped in a state of local energy minimum.

Consider a parallel-striped wrinkle pattern described by (5.51). The average strain energy per unit area of the film consists of three parts [33], namely,

$$\overline{U}(A, k, \theta) = \overline{U}_C + \overline{U}_B + \overline{U}_S, \tag{5.56}$$

where

$$\overline{U}_C = \frac{1}{4}\sigma_\theta h k^2 A^2 + \frac{3}{64}\overline{E}_\theta h k^4 A^4, \tag{5.57}$$

$$\overline{U}_B = \frac{1}{48}\overline{E}_\theta h^3 k^4 A^2, \tag{5.58}$$

$$\overline{U}_S = \frac{1 - \nu_s}{2(1 - 2\nu_s)} \frac{\mu_R}{H} A^2. \tag{5.59}$$

The first two parts, $\overline{U}_C$ and $\overline{U}_B$, are associated with the in-plane deformation and bending of the elastic film, respectively, whereas $\overline{U}_S$ is the reversible elastic strain energy stored in the viscoelastic layer.

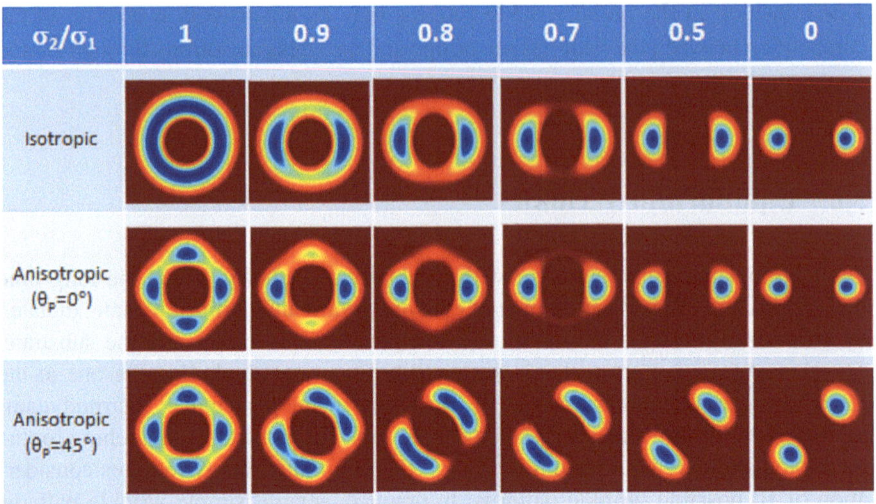

Fig. 5.6 Energy spectra of parallel wrinkles at the equilibrium state for isotropic and anisotropic elastic films. The major principal stress is $\sigma_1/C_{11} = -0.003$

For an arbitrary wave vector, minimizing the total energy gives the equilibrium amplitude as a function of k and θ:

$$A_e(k,\theta) = \frac{2\sqrt{6}}{3k}\left[-\frac{\sigma_\theta}{\overline{E}_\theta} - \frac{(kh)^2}{12} - \frac{2(1-\nu_s)}{1-2\nu_s}\frac{\mu_R}{\overline{E}_\theta}\frac{1}{k^2Hh}\right]^{1/2}. \tag{5.60}$$

Correspondingly, the total strain energy at the equilibrium state is also a function of k and θ, i.e., $\overline{U}_e(k,\theta) = \overline{U}(A_e,k,\theta)$. The spectra of the energy, $\overline{U}_e(k,\theta)$, are plotted in Fig. 5.6 for both isotropic and anisotropic elastic films under various residual stress states. Similar to Fig. 5.5, contours of the normalized strain energy, $\overline{U}_e/(C_{11}h)$, are plotted in the plane spanning the x and y components of the normalized wave vector; only negative energy values are plotted.

For an isotropic elastic film, the energy contours are concentric circles under an equi-biaxial stress $(\sigma_2/\sigma_1 = 1)$. The strain energy minimizes on a circle of a particular radius, with no favored direction due to rotational symmetry of the isotropic system. Once the ratio between the two principal stresses deviates from 1, the rotational symmetry is broken and the energy spectrum has two minima symmetrically located on the axis parallel to the direction of the major principal stress. Hence, an energetically favored wrinkle pattern emerges with parallel stripes perpendicular to the direction of the major principal stress. However, it should be noted that, since only parallel-striped wrinkle patterns are considered in the present analysis, the energy minima in the spectra are not necessarily global minima. For example, it has been shown that herringbone and checkerboard wrinkle patterns may have lower energy than the parallel stripe pattern under an equi-biaxial stress state [36, 39].

For a cubic crystal film (e.g., SiGe), the energy spectrum depends on both the material anisotropy and the stress anisotropy. Under an equi-biaxial stress, there exist four energy minima aligned in the two orthogonal crystal directions, [100] and [010]. Thus, an energetically favored equilibrium wrinkle pattern would consist of parallel stripes in two orthogonal directions. When the two principal stresses are different, the energy spectrum depends on the principal direction θ_p. When $\theta_p = 0$, two energy minima are symmetrically located on the axis parallel to the direction of the major principal stress, while the other two minima in the orthogonal direction first become shallower and then disappear. When $\theta_p \neq 0$, as the stress ratio decreases from 1 to 0, the wave vectors of the energy minima first rotate toward the direction of the major principal stress and then merge to form two minima in the same direction. Therefore, different equilibrium wrinkle patterns may emerge as a transition between the orthogonal and the uniaxial patterns.

Furthermore, by minimizing the strain energy, $\overline{U}_e(k, \theta)$, with respect to k for a fixed angle θ, the equilibrium wavelength is obtained as a function of θ:

$$\lambda_e(\theta) = \frac{2\pi}{k_e(\theta)} = \pi h \left[\frac{2(1-2\nu_s)}{3(1-\nu_s)} \frac{H}{h} \frac{\overline{E}_\theta}{\mu_R} \right]^{1/4}. \tag{5.61}$$

Substitution of the equilibrium wavelength into (5.60) gives the equilibrium amplitude for the parallel wrinkles with the angle θ. Further minimization of the strain energy, $\overline{U}_e(k_e, \theta)$, with respect to θ, gives the angle for the parallel wrinkles with the minimum energy. Both the equilibrium wavelength and the angle can be readily determined from the locations of the minima in the energy spectrum shown in Fig. 5.6. In particular, for an isotropic film under an equi-biaxial residual stress, the equilibrium wrinkle wavelength in (5.61) is independent of the angle, and the corresponding equilibrium wrinkle amplitude is

$$A_e = h\sqrt{\frac{2}{3}\left(\frac{\sigma_0}{\sigma_c} - 1\right)}, \tag{5.62}$$

where σ_c is the critical stress given in (5.50). Recall that we have assumed a thin viscoelastic layer and $\nu_s < 0.5$. For a thick elastic substrate or $\nu_s = 0.5$, the equilibrium wrinkle wavelength and amplitude take different forms [37].

It is noted that, although the energy spectra in Fig. 5.6 appear similar to the initial growth spectra in Fig. 5.5, the locations for the energy minima are different from those of the maximum growth rates. While the fastest growing wavelength depends on the residual stress in the film, the equilibrium wrinkle wavelength is independent of the stress. In fact, the equilibrium wavelength is always greater than the fastest growing wavelength at the initial stage. Consequently, the wrinkle wavelength increases as the wrinkles evolve, a coarsening phenomenon that has been observed in experiments [19, 52]. A scaling analysis of coarsening is presented next in Sect. 5.7, which is confirmed by numerical simulations in Sect. 5.9.

5.7 Coarsening of Wrinkles: A Scaling Analysis

As the wrinkle amplitude grows to be comparable to the film thickness, the residual stress in the film is partly relaxed, and the nonlinear effect of large deflection must be accounted for. It was found that there exists a kinetically constrained wrinkle state for each unstable wavelength [42]. In the neighborhood of such a state, the wrinkling process is very slow (i.e., $\partial w/\partial t \approx 0$), despite the fact that the film remains energetically unstable. For a particular wavelength ($\lambda > \lambda_c$), setting $\partial w/\partial t = 0$ and $R = 0$ in (5.43) leads to a spatially uniform stress in the film

$$\sigma_k = -\frac{4\pi^2 K}{\lambda^2 Fh} = -\frac{\pi^2 h^2 E_f}{3(1-\nu_f^2)\lambda^2}. \tag{5.63}$$

Correspondingly, the wrinkle amplitude is given by [42]

$$A_k = h\sqrt{\frac{1}{3}\left(\frac{\lambda^2}{\lambda_c^2} - 1\right)}. \tag{5.64}$$

Therefore, the film stress can be continuously relaxed by increasing the wrinkle wavelength λ, i.e., coarsening. Meanwhile, the wrinkle amplitude grows.

Assume that the film stays in the neighborhood of the kinetically constrained state during coarsening, with both the wrinkle wavelength and the amplitude evolving nonlinearly. At this stage, the spatial distribution of the stress in the film is nearly uniform, approximately given by (5.63) as a function of the wrinkle wavelength. By comparing the first two terms on the right-hand side of (5.43), the time scale for this stage becomes

$$\tau_2 = \frac{K}{(Fh\sigma_k)^2} = \frac{\lambda^4}{16\pi^4 K}. \tag{5.65}$$

Thus, the wrinkle wavelength scales with the time as

$$\lambda \sim (Kt)^{1/4}. \tag{5.66}$$

The corresponding wrinkle amplitude can be obtained from (5.64). For $\lambda >> \lambda_c$, the same scaling law applies for the growth of wrinkle amplitude, i.e., $A \sim (Kt)^{1/4}$.

Remarkably, a molecular dynamics (MD) simulation of buckling molecular chains showed a very similar scaling with the wavelength and the amplitude growing as a power of time: $\lambda \sim t^n$ and $A \sim t^\beta$, where both the exponents nearly equal to 0.26 [53]. MD simulations of compressed solid membrane showed a similar coarsening dynamics with slightly larger exponents: $n = 0.28$ and $\beta = 0.29$ [54].

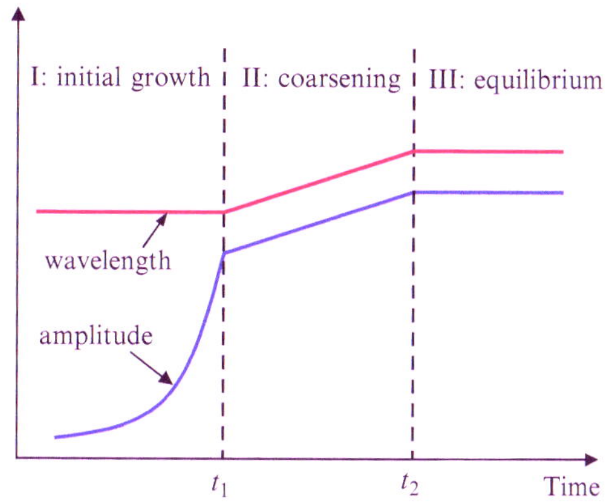

Fig. 5.7 Schematic of the three-stage evolution of wrinkle wavelength and amplitude (both in log-log scale)

It has been noted that the nature of such dynamics is analogous to phase ordering phenomena such as spinodal decomposition [19].

Figure 5.7 schematically illustrates the three stages of wrinkling predicted for the wrinkle wavelength and amplitude: initial growth, coarsening, and equilibrium. The transition points can be determined approximately using the scaling laws [32]. First, the coarsening starts when the wrinkle reaches the kinetically constrained state of the fastest growing mode, which gives the first transition point:

$$t_1 = 4\tau_1 \ln\left(\frac{h}{\sqrt{3}A_0}\right), \tag{5.67}$$

where A_0 is the initial amplitude. Following the power law scaling for the wavelength coarsening, the second transition point is approximately

$$t_2 = t_1 \left(\frac{\lambda_e}{\lambda_m}\right)^4 = \frac{1}{R} \ln\left(\frac{h}{\sqrt{3}A_0}\right). \tag{5.68}$$

Therefore, the first transition time is proportional to the time scale of initial growth, which inversely scales with the square of the residual stress in the film, and the second transition time scales with the relaxation time $(1/R)$ of the viscoelastic layer, independent of the stress.

It should be noted that the above scaling analysis is based on the model assuming that the thickness of the viscoelastic layer is small compared to the wrinkle wavelength and that the viscoelastic layer is compressible (i.e., $\nu_s < 0.5$). Following the same approach, different scaling can be derived for the cases with incompressible thin layers or thick substrates [32]. For an incompressible, thin layer ($\nu_s = 0.5$), the evolution equation takes a different form (5.30), and the coarsening of the wrinkle wavelength scales as $\lambda \sim t^{1/6}$. For a thick viscoelastic substrate ($H/h \to \infty$), the scaling for coarsening is obtained as $\lambda \sim t^{1/3}$. It is thus speculated that, for a viscoelastic layer of finite thickness, the exponent of the power-law scaling for coarsening of wrinkles should be between 1/4 and 1/3 for a compressible layer and between 1/6 and 1/3 for an incompressible layer. Remarkably, a similar transition was noted for the phase separation kinetics of polymer blend films, where the coarsening exponent changes from 1 for thick films (100 nm) to a value near 0.5 for thin films (20 nm) [55].

5.8 A Spectral Method for Numerical Simulations

The nonlinear evolution equations of wrinkling in (5.41) and (5.42) can be solved numerically by a spectral method to simulate evolution of wrinkle patterns [32]. For convenience, the evolution equations are normalized by scaling the lengths, time, and stress components with the film thickness (h), the characteristic time for viscous flow (η_s/μ_f), and the shear modulus (μ_f), respectively. In addition, the linear and nonlinear terms are separated so that Fourier transform of the evolution equations takes the form

$$\frac{\partial \hat{w}}{\partial t} = B\hat{w} + \phi, \tag{5.69}$$

$$\frac{\partial \hat{u}_1}{\partial t} = C_1\hat{u}_1 + D_1\hat{u}_2 + \phi_1, \tag{5.70}$$

$$\frac{\partial \hat{u}_2}{\partial t} = C_2\hat{u}_1 + D_2\hat{u}_2 + \phi_2, \tag{5.71}$$

where a hat over a variable designates its Fourier transform with respect to the in-plane coordinates, and ϕ, ϕ_1, and ϕ_2 represent the nonlinear terms.

A semi-implicit algorithm is employed to integrate (5.69)–(5.71) over time. First, using a backward finite difference scheme for the linear part and a forward scheme for the nonlinear part, (5.69) is integrated point by point in the Fourier space as

$$\hat{w}^{(n+1)} = \frac{\hat{w}^{(n)} + \phi^{(n)}\Delta t}{1 - B\Delta t}, \tag{5.72}$$

where Δt is the time step. The nonlinear terms are calculated in the real space and transformed to the Fourier space numerically by Fast Fourier transform (FFT).

Similarly, (5.70) and (5.71) are integrated as

$$\hat{u}_1^{(n+1)} = \frac{(1 - D_2\Delta t)\left(\hat{u}_1^{(n)} + \phi_1^{(n)}\Delta t\right) + D_1\Delta t\left(\hat{u}_2^{(n)} + \phi_2^{(n)}\Delta t\right)}{(1 - C_1\Delta t)(1 - D_2\Delta t) - C_2D_1(\Delta t)^2}, \tag{5.73}$$

$$\hat{u}_2^{(n+1)} = \frac{(1 - C_1\Delta t)\left(\hat{u}_2^{(n)} + \phi_2^{(n)}\Delta t\right) + C_2\Delta t\left(\hat{u}_1^{(n)} + \phi_1^{(n)}\Delta t\right)}{(1 - C_1\Delta t)(1 - D_2\Delta t) - C_2 D_1 (\Delta t)^2}. \tag{5.74}$$

The advantage of the spectral method is that it resolves the challenges in the numerical simulations by converting the high-order spatial differentiations in the real space into algebraic multiplications in the reciprocal Fourier space, utilizing the efficiency of FFT and its inverse to communicate between the two spaces. Similar methods have been used for simulations of other evolution systems [56–58].

Numerically, a square computational cell in the $x - y$ plane with a periodic boundary condition is used. The size of the computational cell is selected such that it is sufficiently large compared to the longest wrinkle wavelengths (i.e., the equilibrium wavelength λ_e), thus minimizing the effect of cell size on the numerical results. The cell is discretized into grids in both the x- and y-directions with a sufficient resolution in space to resolve the shortest wavelength of interest (i.e., the critical wavelength λ_c). Moreover, the semi-implicit algorithm for time integration is conditionally stable. To insure numerical stability and convergence, the time step Δt (normalized by η_s/μ_f) must be sufficiently small.

5.9 Numerical Simulations of Wrinkle Patterns

In this section, results from numerical simulations are presented to show evolution of diverse wrinkle patterns. For quantitative characterization of the wrinkle patterns, which may be ordered or disordered, two quantities are calculated. First, the wrinkle amplitude is evaluated by a root-mean-square (RMS) of the deflection, namely,

$$\mathrm{RMS}(t) = \sqrt{\frac{\sum w(m,n,t)^2}{N^2}}, \tag{5.75}$$

where $w(m,n,t)$ is the deflection of the grid point (m,n) at time t and N is the number of grid points along each side of the computational cell. Second, the average wrinkle wavelength is evaluated by

$$\bar{\lambda}(t) = \frac{2\pi}{\bar{k}(t)}, \tag{5.76}$$

where

$$\bar{k}(t) = \sqrt{\frac{\sum |\hat{w}(m,n,t)|^2 k(m,n)^2}{\sum |\hat{w}(m,n,t)|^2}}. \tag{5.77}$$

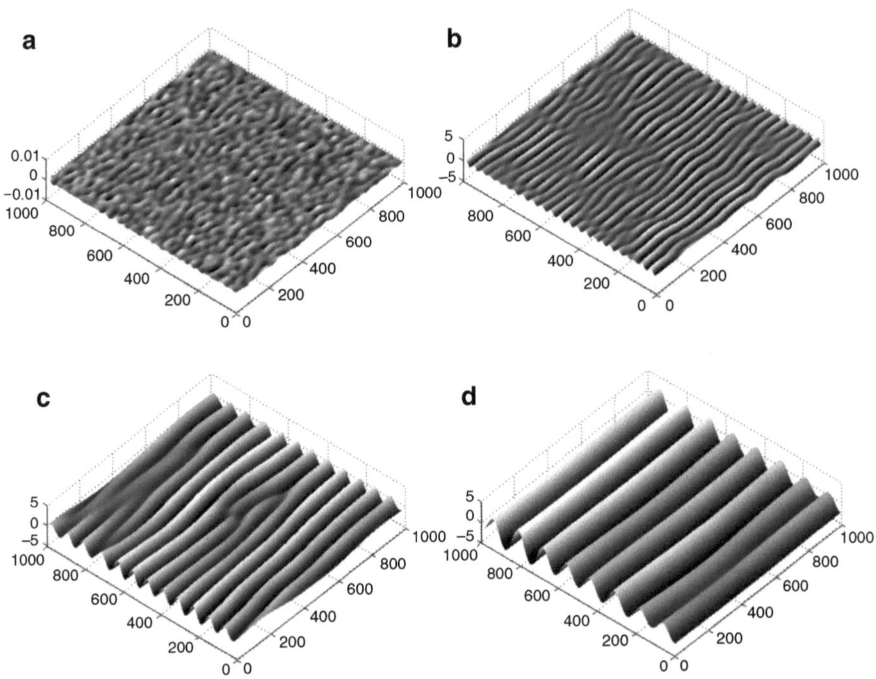

Fig. 5.8 A simulated evolution sequence of wrinkles in a uniaxially stressed film. (**a**) $t = 10^4$: $\bar{\lambda} = 46.2$ and RMS $= 0.00033$; (**b**) $t = 10^5$: $\bar{\lambda} = 44.4$ and RMS $= 0.313$; (**c**) $t = 10^6$: $\bar{\lambda} = 77.8$ and RMS $= 0.947$; (**d**) $t = 10^7$: $\bar{\lambda} = 128.2$ and RMS $= 1.66$

Equation (5.77) may be regarded as a weighted RMS of the wavenumbers in the Fourier space, where the weight $|\hat{w}|^2$ represents the power intensity of the corresponding wavenumber, similar to the power spectra in experiments [19].

5.9.1 Uniaxial Wrinkles (Parallel Stripes)

Figure 5.8 shows an evolution sequence of simulated wrinkle patterns under a uniaxial stress with magnitude $\sigma_0 = 0.01$ (normalized by the shear modulus μ_f). Here the film is isotropic and the modulus ratio μ_R/μ_f is set to zero. The other parameters are $H/h = 10$, $\nu_s = 0.45$, and $\nu_f = 0.3$. The computational cell size is $L/h = 1,000$, with a 128-by-128 grid. The normalized time step is $\Delta t = 100$. To start the simulation, a random perturbation of amplitude 0.001 h is introduced over the entire computational cell.

Under the uniaxial stress, a parallel stripe pattern emerges quickly in the direction perpendicular to the stress. Both the wrinkle amplitude and the wavelength increase with time. The presence of dislocation-type defects in the parallel stripe pattern is evident in Fig. 5.8b, c. The defect density decreases as the wrinkle

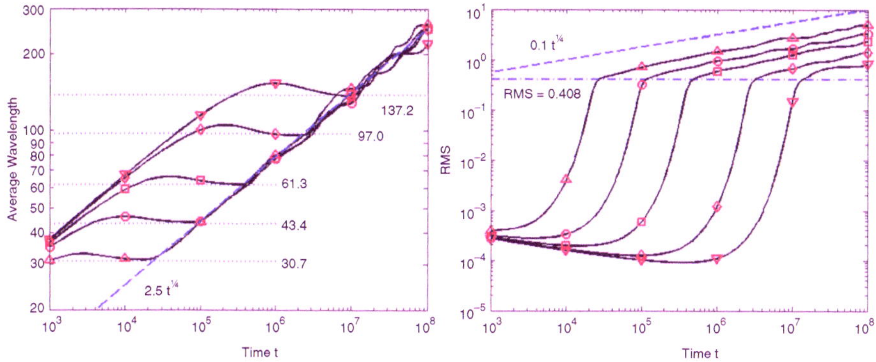

Fig. 5.9 Evolution of average wrinkle wavelength and RMS amplitude under uniaxial stresses ($\mu_R/\mu_f = 0$): $\sigma_0 =$ 0.02 (*open diamond*), 0.01 (*open circle*), 0.005 (*open square*), 0.002 (*open diamond*), 0.001 (*filled inverted triangle*). The *horizontal dotted lines* in (**a**) indicate the fastest growing wavelengths predicted by the linear perturbation analysis. The *dot-dashed line* in (**b**) corresponds to the stress-independent RMS at the kinetically constrained state of the fastest growing mode

evolves. Similar results are obtained for various stress magnitudes. Figure 5.9 plots the evolution of the average wrinkle wavelength and the RMS amplitude. The average wavelength quickly reaches a plateau at a level depending on the residual stress, corresponding to the fastest growing mode predicted by (5.48) for the initial growth. Meanwhile, the RMS amplitude increases exponentially, as predicted by (5.49). The wavelength starts to increase (i.e., coarsening) when the RMS reaches a critical level (~0.4), nearly independent of the residual stress. As indicated by the horizontal dashed line in Fig. 5.9b, the RMS at the kinetically constrained state of wavelength $\lambda = \lambda_m$ is 0.408, which is close to the transition points for all stresses. Therefore, the first transition time in (5.67) is a reasonable estimate for the onset of coarsening.

During coarsening, the wavelength follows a straight line with a slope $1/4$ in the log-log plot as shown in Fig. 5.9a, in good agreement with the scaling analysis in (5.66). Interestingly, in spite of the different residual stresses and different transition points, the wrinkle wavelength follows essentially the same path of coarsening. On the other hand, the wrinkle amplitude grows with the same scaling (after a short transition period) but with different magnitudes for different residual stresses. Since no elastic equilibrium state exists for the present case ($\mu_R/\mu_f = 0$), the coarsening process continues until the simulation is stopped at $t = 10^8$.

The substrate elasticity has two effects on the wrinkling process. First, it can stabilize the film under small stresses so that the film remains flat with no wrinkles at all. The critical stress is given by (5.50). The second effect is that, when the residual stress is high enough to cause wrinkling, the elastic limit of the substrate will eventually stabilize the wrinkle pattern. For parallel-stripe wrinkles, the equilibrium wrinkle wavelength and amplitude are analytically predicted in (5.61) and (5.62), respectively. Figure 5.10 plots evolution of the average wrinkle wavelength

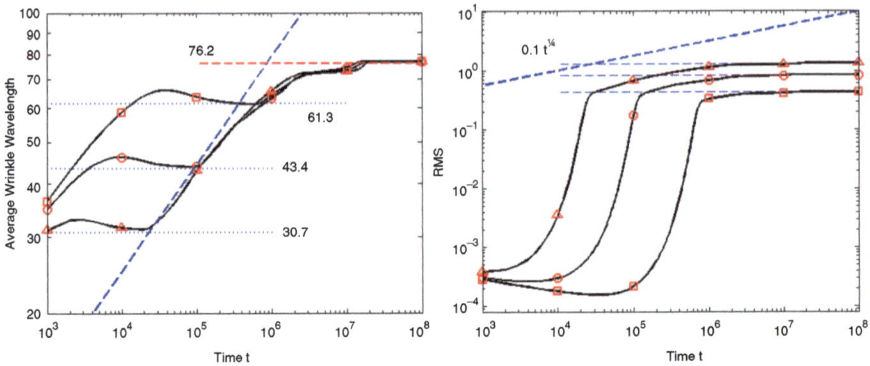

Fig. 5.10 Evolution of average wrinkle wavelength and RMS amplitude under uniaxial stresses: $\sigma_0 =$ 0.02 (*open triangle*), 0.01 (*open circle*), 0.005 (*open square*), with a modulus ratio $\mu_R/\mu_f = 10^{-5}$. The equilibrium wavelength is indicated by the *horizontal dashed line* in (**a**), and the equilibrium amplitudes are indicated by the *horizontal dashed lines* in (**b**)

and the RMS amplitude from numerical simulations under uniaxial stresses and with a modulus ratio, $\mu_R/\mu_f = 10^{-5}$. The critical stress in this case is $\sigma_c = 0.00324$. When $\sigma_0 < \sigma_c$, no wrinkle grows. For $\sigma_0 > \sigma_c$, the initial growth is similar to that in Fig. 5.9, but with a slower growth rate as predicted by the linear perturbation analysis [31]. The coarsening, however, is influenced by the presence of an equilibrium state. The wrinkle wavelength increases and eventually reaches a plateau, in agreement with the predicted equilibrium wavelength, $\lambda_e = 76.2$ in this case. At the same time, the RMS amplitude approaches a plateau in agreement with the predicted equilibrium amplitude. While the equilibrium wavelength is independent of the residual stress, the equilibrium amplitude increases as the stress magnitude increases. Apparently, the power-law scaling for coarsening becomes less predictive as the wrinkles approach the equilibrium state. As a result, the second transition time predicted by (5.68) typically underestimates the time to reach the equilibrium state.

5.9.2 Isotropic Wrinkles (Labyrinths)

Figure 5.11 shows an evolution sequence of simulated wrinkle patterns under an equi-biaxial stress of magnitude $\sigma_0/\mu_f = 0.01$ with $\mu_R/\mu_f = 0$. Differing from the parallel stripe patterns under uniaxial stresses, an isotropically disordered labyrinth pattern emerges under the equi-biaxial stress, due to the rotational symmetry in the isotropic system. Figure 5.12 plots the evolution of the average wrinkle wavelength and the RMS amplitude. Similar to the cases under uniaxial stresses, the average wavelength quickly reaches a plateau, corresponding to the fastest growing mode predicted by the linear analysis, and the RMS increases exponentially during the initial growth stage. The RMS at the transition point from initial growth to

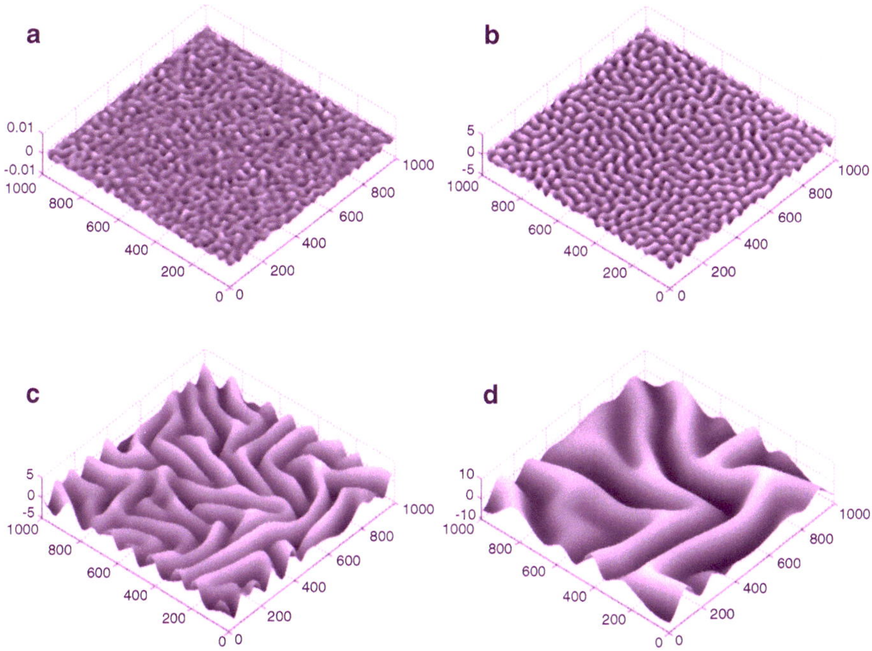

Fig. 5.11 A simulated evolution sequence of wrinkles in an equi-biaxially stressed film. (**a**) $t = 10^4$: $\bar{\lambda} = 42.8$ and RMS $= 0.00054$; (**b**) $t = 10^5$: $\bar{\lambda} = 45.1$ and RMS $= 0.423$; (**c**) $t = 10^6$: $\bar{\lambda} = 98.4$ and RMS $= 1.50$; (**d**) $t = 10^7$: $\bar{\lambda} = 196.9$ and RMS $= 3.18$

coarsening is about the same too, but the transition process is different for the wrinkle wavelength. The coarsening of the wavelength is faster immediately after it reaches the transition point, giving a slope larger than ¼ in the log-log plot (Fig. 5.12a). The slope then decreases as coarsening continues, eventually approaching ¼, as predicted by the scaling analysis. Such a behavior may be attributed to the disorder in the labyrinth wrinkle pattern. At the early stage, the pattern is highly disordered with relatively short stripes in random directions, for which the coarsening is not well described by the scaling law for parallel stripe patterns. Later on, while the wrinkle pattern remains disordered, it consists of long stripes coarsening in a similar manner as parallel stripes. Consequently, the coarsening paths under equi-biaxial stresses are different for different residual stresses, and they are different from those under uniaxial stresses.

Using the length scale L_1 and the time scale τ_1 as defined in (5.45) and (5.46), respectively, Fig. 5.13 replots the evolution of wrinkle wavelength from the numerical simulations (Figs. 5.9a and 5.12a, with $\mu_R/\mu_f = 0$). Remarkably, the evolution paths for different stress magnitudes collapse onto one for uniaxial stresses and another for equi-biaxial stresses. At the early stage, the film selects the fastest growing wrinkle wavelength, which is the same for both uniaxial and equi-biaxial stresses. Interestingly, the process of wavelength selection at the early stage seems

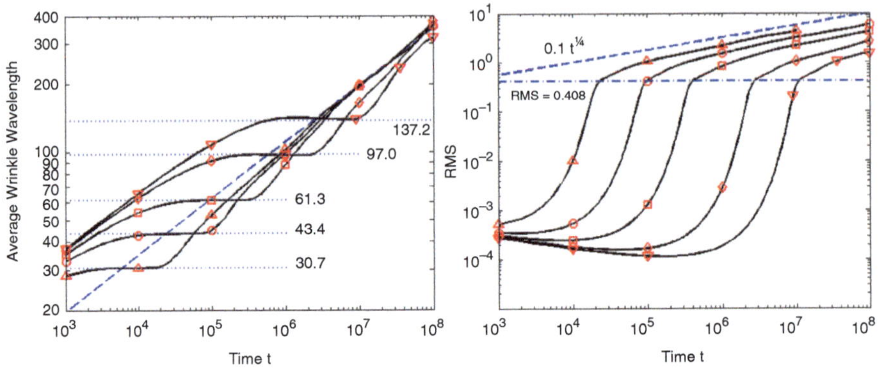

Fig. 5.12 Evolution of average wrinkle wavelength and RMS amplitude under equi-biaxial stresses: $\sigma_0 =$ 0.02 (*open triangle*), 0.01 (*open circle*), 0.005 (*open square*), 0.002 (*open diamond*), 0.001 (*filled inverted triangle*), with $\mu_R/\mu_f = 0$

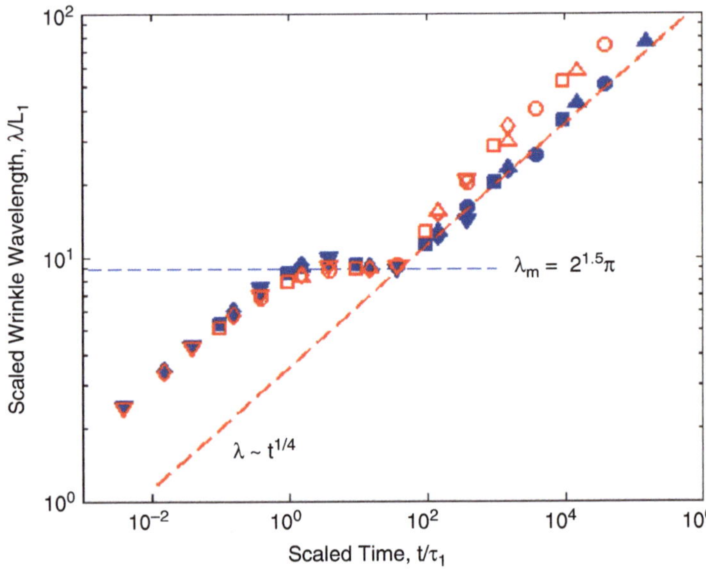

Fig. 5.13 Evolution of wrinkle wavelength, scaled by the length L_1 and the time τ_1, under different uniaxial stresses (*solid symbols*) and equi-biaxial stresses (*open symbols*)

to follow the same scaling law as coarsening, i.e., $\lambda \sim t^{1/4}$. This process, however, may depend on the initial perturbation, and is typically too short to be observed in experiments. The two paths are slightly different in the transition to coarsening. Under uniaxial stresses, the coarsening process is well described by the power-law scaling. Under equi-biaxial stresses, however, it undergoes a transition stage with a faster coarsening rate before it approaches the same power law. This is consistent

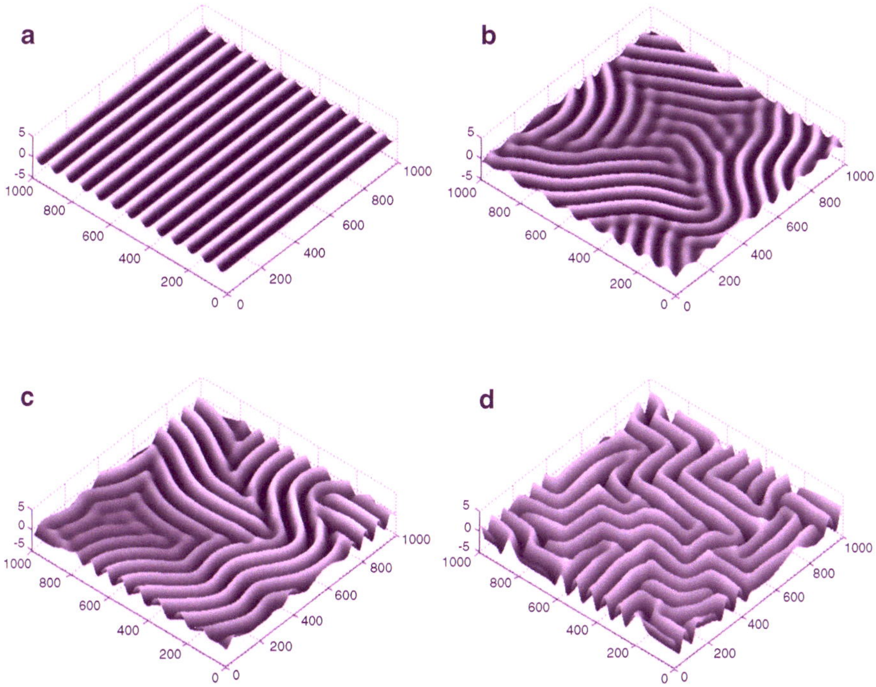

Fig. 5.14 Nearly equilibrium wrinkle patterns: (**a**) under a uniaxial stress, $\sigma_0 = 0.005$; (**b**)–(**d**) under equi-biaxial stresses, $\sigma_0 = 0.005$, 0.01, and 0.02

with the MD simulations of buckling molecular chains [53] and membranes [54], with the chains under uniaxial stresses and the membranes under equi-biaxial stresses.

By adding substrate elasticity ($\mu_R/\mu_f>0$), the average wrinkle wavelength and the RMS amplitude eventually saturate at the equilibrium state [32], similar to the uniaxial wrinkles as shown in Fig. 5.10. Figure 5.14 shows the wrinkle patterns for different stresses at the end of each simulation ($t = 10^8$) when the system has reached a nearly equilibrium state. Under uniaxial stresses (Fig. 5.14a), similar patterns with parallel stripes are obtained for different stress magnitudes, with the same wrinkle wavelength but different amplitudes. Under equi-biaxial stresses (Fig. 5.14b–d), although the average wavelength at equilibrium is independent of the residual stress, the coarsening dynamics leads to different evolution paths and thus different wrinkle patterns. Interestingly, the isotropic wrinkle patterns seem to organize into a domain structure, with parallel stripes in each domain, while the domain size decreases as the stress magnitude increases. The details of the equilibrium wrinkle pattern under equi-biaxial stresses may depend on the initial random perturbation, but the average wrinkle wavelength and the RMS amplitude are insensitive to the initial perturbation. Thus, the apparently chaotic pattern may be characterized by the two deterministic quantities.

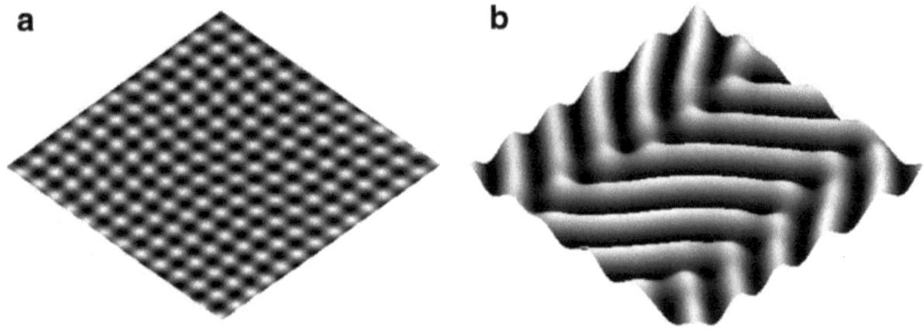

Fig. 5.15 Ordered wrinkle patterns under equi-biaxial compression: (**a**) Checkerboard pattern; (**b**) Herringbone pattern

In addition to the labyrinth pattern, checkerboard and herringbone patterns have also been considered as the possible equilibrium wrinkle patterns under equi-biaxial stresses [36, 39, 40]. As shown in Fig. 5.15a, by the kinetics approach, a checkerboard pattern is obtained when the equi-biaxial residual stress is slightly above the critical stress for the onset of wrinkling. A transition from the checkerboard pattern to the labyrinth pattern occurs as the stress magnitude increases. Such a transition may be attributed to a geometric origin [37]: a flat film can bend to a cylindrical surface with no stretching, but bend to a spherical cap with severe stretching at the crest. Starting from a flat film with a small random perturbation, wrinkles emerge randomly at the early stage, forming shallow spherical dimples. As the wrinkle amplitude grows, the spherical dimples elongate to form cylindrical ridges. The competition between bending and stretching leads to the transition from spherical dimples to cylindrical ridges, while the overall rotational symmetry is maintained as the cylindrical ridges form an isotropic labyrinth pattern. When the residual stress is low, the equilibrium wrinkle amplitude is small, and the shallow spherical dimples remain at the equilibrium state to form the checkerboard pattern. Such an ordering however may be an artifact of the square-shaped computational cell used in the numerical simulation. In a recent experiment, hexagonally packed dimples have been observed [59, 60], which manifests the rotational symmetry in the isotropic system. Figure 5.15b shows a herringbone pattern, which is macroscopically isotropic. Such a pattern appears to be a special case of the labyrinth pattern, with locally ordered parallel stripes and a parallel domain structure. Alternatively, it may also be considered as a combination of the checkerboard pattern (spherical dimples) and the parallel stripe pattern [61]. It was found that, among the three ordered wrinkle patterns (parallel stripes, checkerboard, and herringbone), the herringbone pattern yields the lowest elastic energy in an equi-biaxially compressed film at relatively large overstresses [36, 39]. Recently, Cai et al. [60] presented an energetic analysis that predicts the transition stress from the checkerboard pattern to the herringbone pattern under equi-biaxial compression. They also suggested that, with a small initial curvature of the film, the hexagonal pattern has a lower elastic energy than the checkerboard pattern at small overstresses.

5.9.3 Transition of Wrinkle Patterns: Secondary Buckling Instability

Apparently, the equilibrium wrinkle patterns in an isotropic elastic film depend sensitively on the residual stress at the reference state. A uniaxial stress results in a parallel-stripe wrinkle pattern, and an isotropically disordered labyrinth pattern often emerges under an equi-biaxial residual stress. In between, when the film is subject to a generally biaxial stress, a transition is expected as the ratio between the two principal stresses (σ_2/σ_1) varies from 0 to 1. In particular, while wrinkles in the parallel stripe pattern effectively relax the compressive stress in one direction, the presence of a compressive stress in the perpendicular direction may induce a secondary buckling instability that leads to the transition of wrinkling patterns [34, 40].

Consider a perturbation to the parallel wrinkles, with the out-of-plane displacement taking the form

$$w(x,y,t) = A\cos\{k_1[x + B\cos(k_2 y)]\}. \tag{5.78}$$

The elastic strain energy of the system consists of the bending and in-plane strain energy in the film and the elastic energy in the substrate [34]:

$$\overline{U}_B = \frac{E_{\mathrm{f}}h^3}{48(1-\nu_{\mathrm{f}}^2)}k_1^2A^2\left(k_1^2 + k_1^2k_2^2B^2 + \frac{1}{2}k_2^4B^2 + \frac{3}{8}k_1^2k_2^4B^4\right), \tag{5.79}$$

$$\begin{aligned}\overline{U}_C = {} & \frac{h}{2E_{\mathrm{f}}}\left(\sigma_1^2 + \sigma_2^2 - 2\nu_{\mathrm{f}}\sigma_1\sigma_2\right) \\ & + \frac{1}{8}hA^2k_1^2(2\sigma_1 + \sigma_2B^2k_2^2) + \frac{3E_f}{64(1-\nu_{\mathrm{f}}^2)}hA^4k_1^4\left(1 + \nu_{\mathrm{f}}B^2k_2^2 + \frac{3}{8}B^4k_2^4\right),\end{aligned} \tag{5.80}$$

$$\overline{U}_S = \frac{1}{2}\frac{1-\nu_{\mathrm{s}}}{(1-2\nu_{\mathrm{s}})}\frac{\mu_{\mathrm{R}}}{H}A^2. \tag{5.81}$$

The total strain energy is

$$\overline{U}_{\mathrm{total}} = \overline{U}_B + \overline{U}_C + \overline{U}_S = \overline{U}(A,k_1) + \varDelta\overline{U}(A,B,k_1,k_2). \tag{5.82}$$

The first term on the right-hand side of (5.82) is the strain energy for parallel wrinkles, for which the equilibrium amplitude and wavelength are obtained in Sect. 5.6. The secondary buckling amplitude (B) and the wavenumber (k_2) are then obtained by minimizing the second term, which is

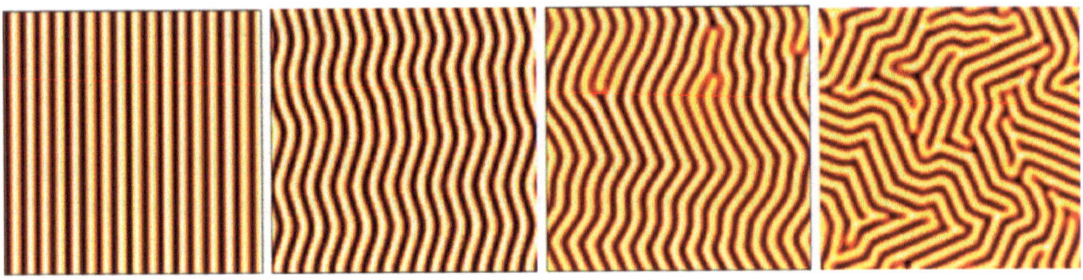

Fig. 5.16 Transition of wrinkle patterns for an isotropic elastic film, from parallel stripes to zigzag and to labyrinth, subject to biaxial residual stresses with $\sigma_1/\bar{E}_f = -0.005$ and $\sigma_2/\bar{E}_f = -0.0028$, -0.0032, -0.0035, -0.005 (from *left* to right)

$$\Delta\overline{U} = \frac{(|\sigma_1| - \sigma_c)^2 h}{36\overline{E}_f}\left[\frac{\overline{E}_f k_2^2 h^2}{|\sigma_1| - \sigma_c} - 12\left(\frac{|\sigma_2| - \sigma_c}{|\sigma_1| - \sigma_c} - \nu_f\right)\right] k_2^2 B^2 + O(B^4). \quad (5.83)$$

The leading term on the right-hand side of (5.83) is quadratic of the amplitude B. When the coefficient is negative, the total energy in (5.82) decreases with the amplitude B, and thus the parallel wrinkles become unstable. The critical condition for the secondary buckling is then predicted as

$$\frac{|\sigma_2| - \sigma_c}{|\sigma_1| - \sigma_c} = \nu_f. \quad (5.84)$$

The same result was obtained by Audoly and Boudaud [40] using a different perturbation function. Apparently, Poisson's ratio of the film plays an important role for the secondary buckling.

By the kinetics approach, evolution of wrinkle patterns under biaxial stresses is simulated. Figure 5.16 shows the near-equilibrium wrinkle patterns obtained from numerical simulations. The major principal stress is identical in all simulations ($\sigma_1 = -0.005\overline{E}_f$) while σ_2 varies between zero and σ_1. The following parameters were used: $\mu_R/\overline{E}_f = 10^{-5}$, $\nu_f = 0.3$, $\nu_s = 0.45$, and $H/h = 10$. The critical stress for the onset of wrinkling (5.50) in this case is $\sigma_c = 0.0019\overline{E}_f$, and the transition stress predicted by (5.84) is $\sigma_2 = -0.00284\overline{E}_f$. Therefore, when $0 > \sigma_2 > -0.00284\overline{E}_f$, parallel wrinkles form and they are stable. When $-0.00284\overline{E}_f > \sigma_2 > \sigma_1$, parallel wrinkles become unstable and zigzag wrinkle patterns emerge. The wavelength of the zigzag and its amplitude may be predicted by energy minimization including the high-order terms in (5.83). Due to the relatively long zigzag wavelength, a much larger computational cell is needed to eliminate the effect of cell size on the simulated zigzag patterns. Nevertheless, the numerical simulations clearly demonstrate the transition from the parallel stripe pattern to the zigzag pattern. As σ_2 approaches σ_1, the zigzag wrinkle pattern becomes increasingly disordered with defects. When $\sigma_2 = \sigma_1$, the wrinkles become isotropically disordered in a labyrinth pattern.

Transition of wrinkle patterns has been observed experimentally [11, 62, 63]. By sequential stretch and release, Lin and Yang [63] reported reversible transition of wrinkle patterns from parallel stripes to zigzag and to highly ordered herringbone patterns. In their experiments, a square-shaped polydimethylsiloxane (PDMS) strip was first stretched equi-biaxially and then treated with oxygen plasma to form a thin oxide layer at the surface. Next, the pre-stretch was released either sequentially in the x- and y-directions or simultaneously in both directions. By the sequential release, they observed a transition in the wrinkle pattern from parallel stripes to zigzag, which qualitatively agrees with the secondary buckling analysis and the numerical simulations shown in Fig. 5.16. Interestingly, a highly ordered herringbone pattern was obtained at the end of the sequential release when the oxide layer was under equi-biaxial compression. On the other hand, by simultaneous release, a disordered labyrinth pattern was obtained at the same final state of equi-biaxial compression. Apparently, the wrinkle pattern depends on the loading path in these experiments. The well-ordered parallel wrinkles formed during the first stage of sequential release seem to have placed seeding for the ordering of subsequent wrinkle patterns. Without any seeding, wrinkles grow in all directions during the simultaneous release, resulting in a disordered labyrinth pattern. Indeed, to obtain the herringbone pattern by numerical simulations as shown in Fig. 5.15b, the residual stress in the film was assumed to be time dependent to simulate a sequential loading so that parallel wrinkles formed first to be followed by zigzag and eventually herringbone patterns. Otherwise, disordered labyrinth patterns emerge from a random initial perturbation under equi-biaxial compression, as shown in Figs. 5.11 and 5.14. The path-dependent wrinkle patterns suggest that there may exist more than one equilibrium states (ordered and disordered) although some of them may be thermodynamically metastable. As such, by controlling the loading path, ordered wrinkle patterns can be achieved.

5.9.4 *Wrinkle Patterns of Anisotropic Crystal Films*

Most of the theoretical studies on wrinkling have assumed isotropic, linear elastic properties for the film. Several experimental studies have reported wrinkling of anisotropic crystal films. Hobart et al. [51] observed wrinkling of single-crystal silicon-germanium (SiGe) alloy films on a glass layer when annealed at an elevated temperature, where SiGe is a cubic crystal with anisotropic elastic properties. More detailed experiments [52] showed that the SiGe film preferentially wrinkles in two orthogonal directions, closely aligned with the <100> axes of the crystal film. Similar orthogonal patterns were observed for a SiGe/oxide film stack [64]. On the other hand, Choi et al. [9] observed zigzag wrinkle patterns of single-crystal Si films bonded to a PDMS substrate, where the jog angle of the zigzag was close to 90°, although no alignment with the crystal axes was reported. By the kinetics approach, coupled with the equilibrium equations for an anisotropic thin film (Sect. 5.2), wrinkle patterns of anisotropic crystal films can be simulated under

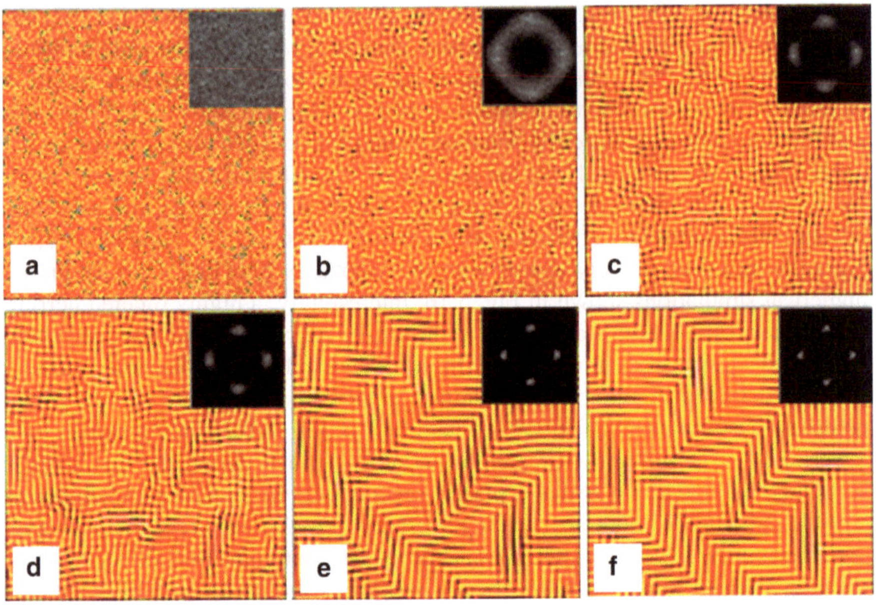

Fig. 5.17 A simulated evolution sequence of the wrinkle pattern for a SiGe crystal film under an equi-biaxial residual stress $(\sigma_1 = \sigma_2 = -0.003C_{11})$: (**a**) $t = 0$, RMS $= 0.0057$, $\bar{\lambda} = 38.83$; (**b**) $t = 10^5$, RMS $= 0.0165$, $\bar{\lambda} = 44.77$; (**c**) $t = 5 \times 10^5$, RMS $= 0.4185$, $\bar{\lambda} = 47.06$; (**d**) $t = 10^6$, RMS $= 0.4676$, $\bar{\lambda} = 50.75$; (**e**) $t = 10^7$, RMS $= 0.5876$, $\bar{\lambda} = 56.43$; (**f**) $t = 10^8$, RMS $= 0.5918$, $\bar{\lambda} = 56.63$

various stress states [33]. In particular, the effects of anisotropic elastic property of a cubic crystal film on wrinkle patterns at both the initial growth stage and the equilibrium state have been elucidated by the analytical solutions in Sects. 5.5 and 5.6, respectively. By the spectral method (Sect. 5.8), Fig. 5.17 shows an evolution sequence of the simulated wrinkle pattern for a cubic crystal film (SiGe) under an equi-biaxial compression $(\sigma_1 = \sigma_2 = -0.003C_{11})$. Here, a square computational cell of size $L = 2,000h$ is discretized into a 128-by-128 grid, with periodic boundary conditions. A random perturbation of amplitude $0.01h$ was introduced as the initial perturbation from the reference state. The deflection, $w(x, y, t)$, is normalized by the film thickness h and plotted as contours in the x–y plane; the time is normalized by the time scale $\tau = \eta/C_{11}$. The insets in Fig. 5.17 show contours in the Fourier space as the Fourier transform of the corresponding wrinkle patterns.

As shown in Fig. 5.17, the initial perturbation at $t = 0$ is featureless, with a small roughness (RMS $= 0.0057$). At $t = 10^5$, the RMS amplitude has grown significantly, and the Fourier transform takes a shape similar to the growth-rate spectrum shown in Fig. 5.5 for the anisotropic film under an equi-biaxial stress. At the early stage, many Fourier components are growing simultaneously, resulting in a disordered wrinkle pattern. The average wrinkle wavelength $(\bar{\lambda} = 44.77)$ is close to the fastest growing wavelength ($\lambda_{\theta m} = 43.23$ for $\theta = 0$) as predicted by the linear analysis. At $t = 5 \times 10^5$, the Fourier components with the fastest growth rate start to dominate, and the wrinkles become increasingly aligned in the two orthogonal directions of the crystal, [100] and [010]. At $t = 10^6$, the wrinkle pattern exhibits a bi-phase domain structure, with parallel stripes locally ordered in one of the two orthogonal directions in each domain. Further evolution of the wrinkle pattern shows two coarsening processes. First, the wavelength of each individual wrinkle stripe increases. As a result, the average wavelength of the wrinkle over the entire area increases, similar to the coarsening process of an isotropic film. Furthermore, the orthogonal bi-phase domain structure of the wrinkle pattern evolves with coarsening of the domain size, as can be seen clearly from $t = 10^6$ to $t = 10^7$. Both the wrinkle wavelength and the domain size seem to saturate after a long-time evolution. It is thus postulated that the viscoelastic evolution process seeks to minimize the total strain energy in the film and the substrate not only by selecting an equilibrium wavelength for individual wrinkle stripes but also by selecting a particular domain size [33]. The present simulation of the wrinkle pattern evolution qualitatively agrees with the experiments by Peterson [52] for a SiGe film on a glass layer at an elevated temperature. The orthogonally ordered wrinkle pattern is also comparable to the wavy structures observed in biaxially stressed silicon membranes on a PDMS substrate [9].

Subject to a generally biaxial stress, the wrinkle pattern depends on the stress anisotropy in addition to the material anisotropy. Similar to the isotropic film, the wrinkle pattern undergoes a transition from the orthogonal pattern under an equi-biaxial stress to the parallel stripe pattern under a uniaxial stress. However, the transition path for the anisotropic film depends on the principal direction of the residual stress [33]. Figure 5.18 shows the wrinkle patterns with two different principal directions. When $\theta_p = 0$, the directions of principal stresses coincide with the <100> crystal axes. In this case, the orthogonal wrinkle pattern remains orthogonal as the stress ratio σ_2/σ_1 slightly deviates from 1, but the bi-phase domain structure changes. Under the equi-biaxial stress ($\sigma_2/\sigma_1 = 1$), the parallel wrinkles are equally distributed in the two orthogonal directions. For $\sigma_2/\sigma_1 = 0.9$, the area decorated with wrinkles in the [100]-direction is greater than the area decorated by wrinkles in the [010]-direction. As the stress ratio σ_2/σ_1 decreases further, the area percentage of the [100] wrinkles increases, and the entire area is covered with the [100] wrinkles when $\sigma_2/\sigma_1 < 0.8$. Thus, the orthogonal wrinkle pattern changes to the uniaxial pattern by eliminating one of the two phases as the stress anisotropy increases. When $\theta_p = 45°$, the directions of the principal stresses are in the <110>-directions of the crystal. As a result, the uniaxial wrinkles are parallel to the <110>-directions, which are absent in the orthogonal pattern under an equi-biaxial stress. As the stress ratio decreases, the directions of the wrinkle stripes first rotate to form zigzag patterns and then merge into the [110]-direction for $\sigma_2/\sigma_1 < 0.8$. Thus, the transition occurs by rotating the wrinkles in both phases toward the major principal direction. Note that the zigzag pattern in transition typically consists of wrinkle stripes in two directions of an oblique angle, consistent with the rotation of

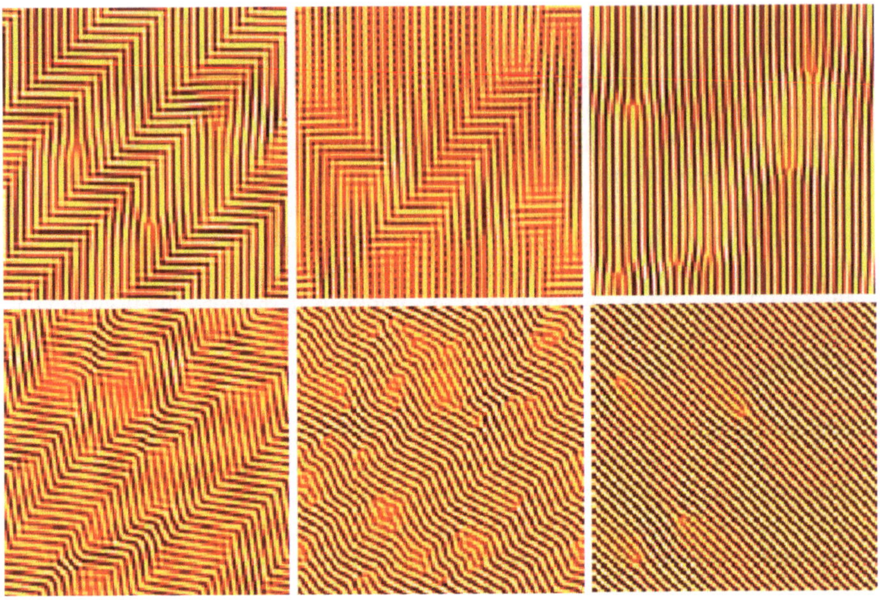

Fig. 5.18 Transition of wrinkle patterns for an anisotropic crystal film, from orthogonal to uniaxial, under biaxial stresses with $\theta_p = 0$ (*upper row*) and $\theta_p = 45°$ (*lower row*). From *left* to *right*: $\sigma_2/\sigma_1 = 0.9$, 0.8, and 0.7

the energy minima in the predicted energy spectra (Fig. 5.6). Therefore, the competition between the stress anisotropy and the material anisotropy further enriches the dynamics of wrinkle patterns, which may be exploited for surface patterning.

5.9.5 Wrinkle Patterns Under Nonuniform Stresses

For a thin film with finite lateral dimensions, the residual stress is typically nonuniform due to stress relaxation along the edges [65–67]. As a result, the wrinkle pattern depends on the distribution of the residual stress in the film, which in turn depends on the shape and size of the film. Two potential applications may be envisaged here. First, one may observe the wrinkle patterns as a visual reflection of the residual stress, thus offering a measure of the underlying stress state. Second, one may design the film geometry to achieve desirable wrinkle patterns for functional applications, such as color control [25] and adhesion [14]. However, predicting the wrinkle patterns under nonuniform residual stresses is often challenging. The energy minimization approach is practically ineffective for this purpose. On the other hand, with slight modification, the kinetics approach has been employed to simulate wrinkle patterns under nonuniform stresses [34]. For simplicity, only isotropic elastic films are considered in this section.

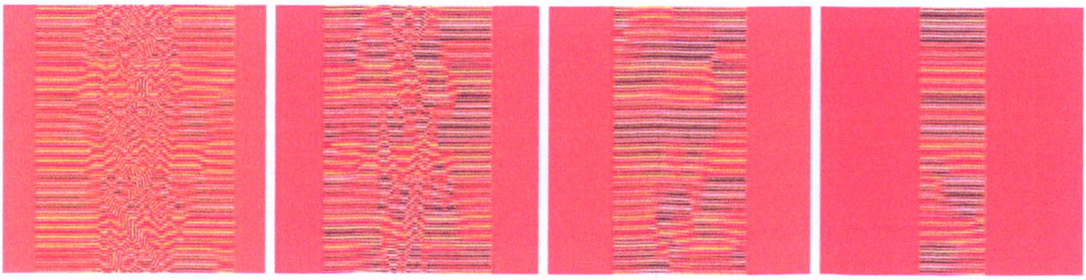

Fig. 5.19 Simulated wrinkle patterns in thin-film ribbons. The ribbon widths are (from *left* to *right*): b/h = 3,750, 3,250, 2,500, and 1,250

First consider a long thin-film ribbon with a finite width b in the x-direction ($-b/2 \leq x \leq b/2$), but infinite in the y-direction. By a shear-lag model [34, 48], the residual stress in the ribbon is obtained as

$$\sigma_{xx}^{\mathrm{R}} = -\sigma_0 \left[1 - \frac{\cosh(x/\lambda)}{\cosh(b/2\lambda)} \right], \tag{5.85}$$

$$\sigma_{yy}^{\mathrm{R}} = -\sigma_0 \left[1 - \nu_{\mathrm{f}} \frac{\cosh(x/\lambda)}{\cosh(b/2\lambda)} \right], \tag{5.86}$$

where σ_0 is the uniform stress magnitude before relaxation, assumed to be equi-biaxial compression, and $\lambda = \sqrt{\overline{E}_{\mathrm{f}} H h / \mu_{\mathrm{R}}}$ is the shear-lag length at the elastic limit of the substrate. The normalized shear-lag length (λ/h) is 1,000 when $\mu_{\mathrm{R}}/\overline{E}_{\mathrm{f}} = 10^{-5}$ and $H/h = 10$. For a narrow ribbon, the stress in the width direction $(\sigma_{xx}^{\mathrm{R}})$ is significantly relaxed over the entire ribbon, while the stress relaxation in the longitudinal direction $(\sigma_{yy}^{\mathrm{R}})$ is relatively small, depending on the Poisson's ratio of the film. Thus, parallel wrinkles are expected to form for the narrow ribbon. As the ribbon width increases, the maximum stress in the x-direction increases. Beyond a critical width, the maximum stress at the center of the ribbon ($x = 0$) becomes greater than the transition stress predicted by (5.84). Consequently, the parallel wrinkles become unstable at the center, where zigzag wrinkle patterns are expected to form. The critical ribbon width for this transition can be predicted by setting the maximum lateral stress equal to the transition stress [34]. Figure 5.19 shows the simulated wrinkle patterns in thin-film ribbons of different widths, by the kinetics approach using the nonuniform residual stresses given by (5.85) and (5.86). For these simulations, the computational cell size is $5{,}000h \times 5{,}000h$, with a 512-by-512 grid. To simulate the free edges, the residual stress is set to be zero outside the ribbon area. As expected, the wrinkles remain uniaxial near the edges. For wide ribbons, the wrinkles become zigzag and disordered at the center. Only when the width of the ribbon is less than a critical value, parallel wrinkles are obtained over the entire ribbon. Similar wrinkle patterns have been observed experimentally [9, 20].

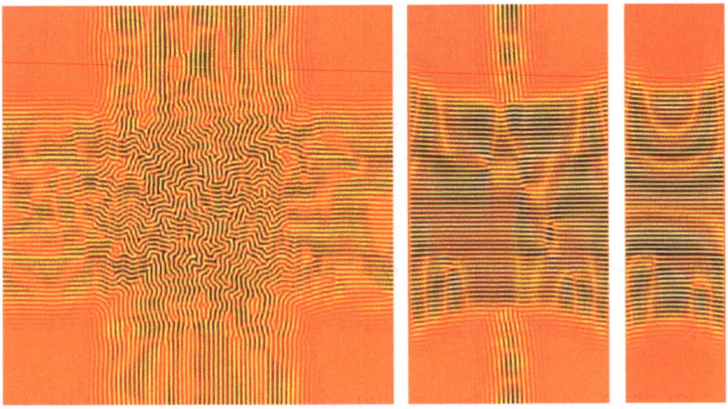

Fig. 5.20 Simulated wrinkle patterns of a square film and two rectangular films, with $b_y/h = 5,000$ and $b_x/h = 5,000$, 2,500, and 1,250

Next consider a rectangular thin film with widths b_x and b_y in the x- and y-directions, respectively. A two-dimensional (2D) shear-lag model may be used to determine the stress distribution in the rectangular film, which however can only be obtained numerically [68]. As an approximation, we extend the 1D solution in (5.85) and (5.86) to 2D as

$$\sigma_{xx}^{\mathrm{R}} = -\sigma_0 \left[1 - \frac{\cosh(x/\lambda)}{\cosh(b_x/2\lambda)} \right] \left[1 - \nu_{\mathrm{f}} \frac{\cosh(y/\lambda)}{\cosh\left(b_y/2\lambda\right)} \right], \tag{5.87}$$

$$\sigma_{yy}^{\mathrm{R}} = -\sigma_0 \left[1 - \nu_{\mathrm{f}} \frac{\cosh(x/\lambda)}{\cosh(b_x/2\lambda)} \right] \left[1 - \frac{\cosh(y/\lambda)}{\cosh\left(b_y/2\lambda\right)} \right]. \tag{5.88}$$

Using the approximate stress distribution, Fig. 5.20 shows the simulated wrinkle patterns of a square film and two rectangular films [34]. For the square film $(b_x = b_y = 5,000h)$, parallel wrinkles develop near the mid-portion of the edges, while the four corners are wrinkle free due to stress relaxation in both directions. The center part of the film shows a fully developed labyrinth pattern. For the rectangular film with $b_x = 2,500h$, predominantly parallel wrinkles extend from left to right except near the top and bottom edges where a small number of wrinkles in the perpendicular direction remain. With $b_x = 1,250h$, only parallel wrinkles appear, with flat regions at the top and bottom. Choi et al. [9] observed similar wrinkle patterns in rectangular Si membranes bonded to a pre-strained PDMS. They also showed wrinkle patterns of other membrane geometries, including circles, ovals, hexagons, and triangles. For some applications, wrinkling may be undesirable. A wrinkle-free film may be achieved if the lateral dimensions are sufficiently small [44, 52, 66].

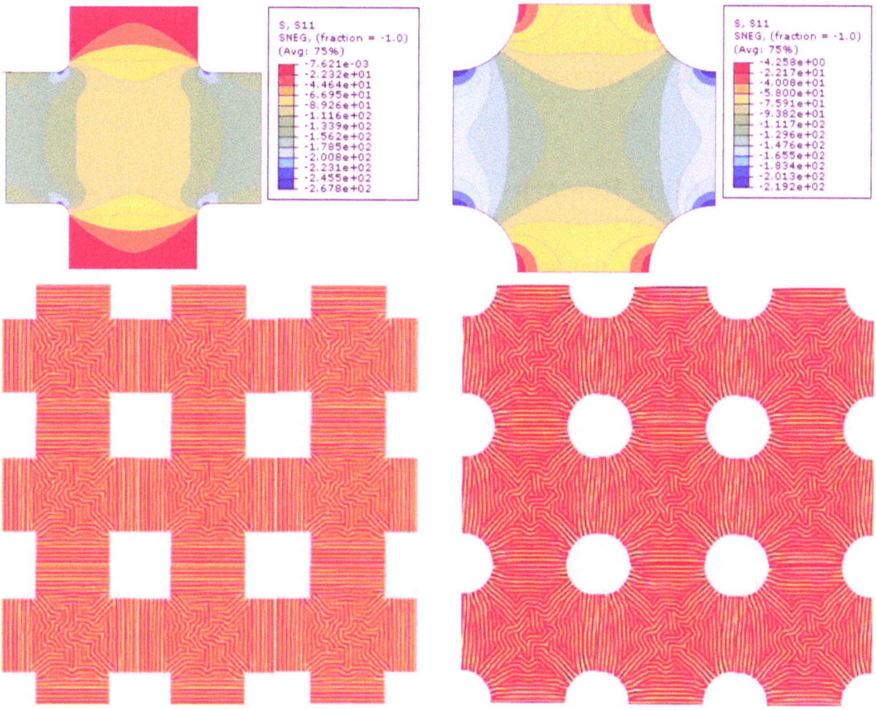

Fig. 5.21 Wrinkling of periodically patterned films. Upper: Distribution of residual stresses in a unit cell obtained from a finite element analysis. Lower: Periodical replication of the simulated wrinkle patterns

To obtain ordered wrinkle patterns, Bowden et al. [10] fabricated bas-relief patterned surfaces and obtained a variety of wrinkle patterns with a thin metal film on an elastomer substrate. With the patterned surface, the residual stress in the metal film is likely nonuniform. However, the stress distribution is often complicated and cannot be predicted analytically. By the finite element method, the distribution of the residual stress in patterned thin films can be obtained numerically, assuming no wrinkles. Figure 5.21 shows the stress distributions for periodic patterns with square and circular holes [34]. The unit cell of each pattern is modeled as a two-dimensional plate on an elastic foundation, and the residual stress is calculated by invoking a mismatch in thermal expansion. Next, with the nonuniform stress distribution, evolution of wrinkle patterns is simulated by the kinetics approach [34]. To compare with the experimental observations, the resulting wrinkle patterns in the unit cell are replicated to produce the periodic wrinkles in a larger scale in Fig. 5.21, showing remarkably similar features as those reported by Bowden et al. [10]. Therefore, it is possible to predict complex wrinkle patterns by combining the kinetics approach with the finite element method.

5.10 Concluding Remarks

In summary, a kinetics approach to wrinkling is presented in this chapter. It is demonstrated that the kinetics approach is capable of simulating evolution of wrinkle patterns under various conditions. Furthermore, it has enabled us to understand the dynamics of coarsening and pattern transition both analytically and numerically. The effects of stress anisotropy, material anisotropy, and nonuniform stress distribution on wrinkle patterns are elucidated by comparing numerical simulations with experimental observations.

Despite significant advances over the last decade in both theory and experiments, recent studies continue to show fascinating wrinkles that stimulate further studies. A few notable examples include concomitant wrinkling and delamination [69, 70], diffusion-induced wrinkling [71, 72], and wrinkle patterns on curved surfaces [73, 74]. In addition to fundamental understanding, several recent reviews have highlighted potential applications of the wrinkling phenomena in stretchable electronics [75], metrology [76], adhesion [77], and micro/nano-fluidics [78].

Acknowledgments The work presented here was carried out in collaboration with S.H. Im. The author gratefully acknowledges fruitful interactions with Z. Suo and C.M. Stafford over the years. This work was supported by National Science Foundation, through Grant No. 0547409.

References

1. Gough GS, Elam CF, De Bruyne ND (1940) The stabilization of a thin sheet by a continuous supporting medium. J R Aeronaut Soc 44:12–43
2. Wan CC (1947) Face buckling and core strength requirements in sandwich construction. J Aeronaut Sci 14:531–539
3. Goodier JN, Neou IM (1951) The evaluation of theoretical critical compression in sandwich plates. J Aeronaut Sci 18:649–657
4. Allen HG (1969) Analysis and design of structural sandwich panels. Pergamon, New York
5. Biot MA (1957) Folding instability of a layered viscoelastic medium under compression. Proc R Soc Lond A 242:444–454
6. Biot MA (1963) Surface instability of rubber in compression. Appl Sci Res A 12:168–182
7. Watanabe M, Shirai H, Hirai T (2002) Wrinkled polypyrrole electrode for electroactive polymer actuators. J Appl Phys 92(8):4631–4637
8. Lacour SP, Wagner S, Huang ZY, Suo Z (2003) Stretchable gold conductors on elastomeric substrates. Appl Phys Lett 82(15):2404–2406
9. Choi WM, Song J, Khang D-Y, Jiang H, Huang Y, Rogers JA (2007) Biaxially stretchable "wavy" silicon nanomembranes. Nano Lett 7:1655–1663

10. Bowden N, Brittain S, Evans AG, Hutchinson JW, Whitesides GM (1998) Spontaneous formation of ordered structures in thin films of metals supported on an elastomeric polymer. Nature 393:146–149
11. Ohzono T, Shimomura M (2004) Ordering of microwrinkle patterns by compressive strain. Phys Rev B 69:132202
12. Chan EP, Crosby AJ (2006) Fabricating microlens arrays by surface wrinkling. Adv Mater 18:3238–3242
13. Harrison C, Stafford CM, Zhang W, Karim A (2004) Sinusoidal phase grating created by a tunably buckled surface. Appl Phys Lett 85:4016–4018
14. Chan EP, Smith EJ, Hayward RC, Crosby AJ (2008) Surface wrinkles for smart adhesion. Adv Mater 20:711–716
15. Stafford CM, Harrison C, Beers KL, Karim A, Amis EJ, Vanlandingham MR, Kim H-C, Volksen W, Miller RD, Simonyi EE (2004) A buckling-based metrology for measuring the elastic moduli of polymeric thin films. Nat Mater 3:545–550
16. Stafford CM, Vogt BD, Harrison C, Julthongpiput D, Huang R (2006) Elastic moduli of ultrathin amorphous polymer films. Macromolecules 39:5095–5099
17. Chan EP, Page KA, Im SH, Patton DL, Huang R, Stafford CM (2009) Viscoelastic properties of confined polymer films measured via thermal wrinkling. Soft Matter 5:4638–4641
18. Volynskii AL, Bazhenov S, Lebedeva OV, Bakeev NF (2000) Mechanical buckling instability of thin coatings deposited on soft polymer substrates. J Mater Sci 35(3):547–554
19. Yoo PJ, Lee HH (2003) Evolution of a stress-driven pattern in thin bilayer films: spinodal wrinkling. Phys Rev Lett 91(15):154502
20. Okayasu T, Zhang HL, Bucknall DG, Briggs GAD (2004) Spontaneous formation of ordered lateral patterns in polymer thin-film structures. Adv Funct Mater 14:1081–1088
21. Kucken M, Newell AC (2004) A model for fingerprint formation. Europhys Lett 68:141–146
22. Volokh KY (2006) Tissue morphogenesis: a surface buckling mechanism. Int J Dev Biol 50:359–365
23. Hilgetag CC, Barbas H (2006) Role of mechanical factors in the morphology of the primate cerebral cortex. PLoS Comput Biol 2:146–159
24. Sultan E, Boudaoud A (2008) The buckling of a swollen thin gel layer bound to a compliant substrate. J Appl Mech 75:051002
25. Xie T, Xiao X, Li J, Wang R (2010) Encoding localized strain history through wrinkle based structural colors. Adv Mater 20:4390–4394
26. Tanaka T, Sun S-T, Hirokawa Y, Katayama S, Kucera J, Hirose Y, Amiya T (1987) Mechanical instability of gels at the phase transition. Nature 325:796–798
27. Gent AN, Cho IS (1999) Surface instabilities in compressed or bent rubber blocks. Rubber Chem Technol 72:253–262
28. Guvendiren M, Yang S, Burdick JA (2009) Swelling-induced surface patterns in hydrogels with gradient crosslinking density. Adv Funct Mater 19:3038–3045
29. Kang MK, Huang R (2010) Swell induced surface instability of confined hydrogel layers on substrates. J Mech Phys Solids 58:1582–1598
30. Huang R (2005) Kinetic wrinkling of an elastic film on a viscoelastic substrate. J Mech Phy Solids 53:63–89
31. Im SH, Huang R (2005) Evolution of wrinkles in elastic-viscoelastic bilayer thin films. J Appl Mech 72:955–961
32. Huang R, Im SH (2006) Dynamics of wrinkle growth and coarsening in stressed thin films. Phys Rev E 74:026214
33. Im SH, Huang R (2008) Wrinkle patterns of anisotropic crystal films on viscoelastic substrates. J Mech Phys Solids 56:3315–3330
34. Im SH (2009) Wrinkling of elastic thin films on compliant substrates. PhD dissertation, University of Texas at Austin
35. Groenewold J (2001) Wrinkling of plates coupled with soft elastic media. Physica A 298:32–45

36. Chen X, Hutchinson JW (2004) Herringbone buckling patterns of compressed thin films on compliant substrates. J Appl Mech 71(5):597–603
37. Huang ZY, Hong W, Suo Z (2005) Nonlinear analyses of wrinkles in a film bonded to a compliant substrate. J Mech Phys Solids 53(9):2101–2118
38. Jiang H, Khang DY, Song J, Sun Y, Huang Y, Rogers JA (2007) Finite deformation mechanics in buckled thin films on compliant supports. Proc Natl Acad Sci 104(40):15607–15612
39. Song J, Jiang H, Choi WM, Khang DY, Huang Y, Rogers JA (2008) An analytical study of two-dimensional buckling of thin films on compliant substrates. J Appl Phys 103:014303
40. Audoly B, Boudaoud A (2008) Buckling of a stiff film bound to a compliant substrate part I: formulation, linear stability of cylindrical patterns, secondary bifurcations. J Mech Phys Solids 56(7):2401–2421
41. Sridhar N, Srolovitz DJ, Suo Z (2001) Kinetics of buckling of a compressed film on a viscous substrate. Appl Phys Lett 78(17):2482–2484
42. Huang R, Suo Z (2002) Wrinkling of a compressed elastic film on a viscous layer. J Appl Phys 91(3):1135–1142
43. Huang R, Suo Z (2002) Instability of a compressed elastic film on a viscous layer. Int J Solids Struct 39(7):1791–1802
44. Liang J, Huang R, Yin H, Sturm JC, Hobart KD, Suo Z (2002) Relaxation of compressed elastic islands on a viscous layer. Acta Mater 50:2933–2944
45. Timoshenko S (1940) Theory of plates and shells. McGraw-Hill, New York
46. Landau LD, Lifshitz EM (1959) Theory of elasticity. Pergamon, London
47. Christensen RM (1982) Theory of viscoelasticity: an introduction. Academic, New York
48. Xia ZC, Hutchinson JW (2000) Crack patterns in thin films. J Mech Phys Solids 48:1107–1131
49. Huang R, Prevost JH, Suo Z (2002) Loss of constraint on fracture in thin film structures due to creep. Acta Mater 50:4137–4148
50. Suo Z, Prévost JH, Liang J (2003) Kinetics of crack initiation and growth in organic-containing integrated structures. J Mech Phys Solids 51:2169–2190
51. Hobart KD, Kub FJ, Fatemi M, Twigg ME, Thompson PE, Kuan TS, Inoki CK (2000) Compliant substrates: a comparative study of the relaxation mechanisms of strained films bonded to high and low viscosity oxides. J Electron Mater 29(7):897–900
52. Peterson RL (2006) Stretching silicon: a uniaxial and biaxial strain generation process and the resulting mobility enhancement in silicon-on-insulator MOSFETs. PhD dissertation, Princeton University
53. Golubovic L, Moldovan D, Peredera A (1998) Dynamics of the Euler buckling instability. Phys Rev Lett 81:3387–3390
54. Moldovan D, Golubovic L (1999) Buckling dynamics of compressed thin sheets (membranes). Phys Rev Lett 82(14):2884–2887
55. Sung L, Karim A, Douglas JF, Han CC (1996) Dimensional crossover in the phase separation kinetics of thin polymer blend films. Phys Rev Lett 76(23):4368–4371
56. Lu W, Suo Z (2001) Dynamics of nanoscale pattern formation of an epitaxial monolayer. J Mech Phys Solids 49(9):1937–1950
57. Golovin AA, Davis SH, Voorhees PW (2003) Self-organization of quantum dots in epitaxially strained solid films. Phys Rev E 68(5):056203
58. Pang Y, Huang R (2006) Nonlinear effect of stress and wetting on surface evolution in epitaxial thin films. Phys Rev B 74:075413
59. Breid D, Crosby AJ (2011) Effect of stress state on wrinkle morphology. Soft Matter 7:4490–4496
60. Cai S, Breid D, Crosby AJ, Suo Z, Hutchinson JW (2011) Periodic patterns and energy states of buckled films on compliant substrates. J Mech Phys Solids 59:1094–1114
61. Kwon SJ, Lee HH (2005) Theoretical analysis of two-dimensional buckling patterns of thin metal-polymer bilayer on the substrate. J Appl Phys 98:063526
62. Ohzono T, Shimomura M (2005) Effect of thermal annealing and compression on the stability of microwrinkle patterns. Phys Rev E 72:025203(R)

63. Lin P-C, Yang S (2007) Spontaneous formation of one-dimensional ripples in transit to highly ordered two-dimensional herringbone structures through sequential and unequal biaxial mechanical stretching. Appl Phys Lett 90:241903
64. Yu CY, Chen PW, Jan SR, Liao MH, Liao K-F, Liu CW (2005) Buckled SiGe layers by the oxidation of SiGe on viscous SiO_2 layers. Appl Phys Lett 86:011909
65. Shield TW, Kim KS (1992) Beam theory models for thin film segments cohesively bonded to an elastic half space. Int J Solids Struct 29:1085–1103
66. Yin H, Huang R, Hobart KD, Suo Z, Kuan TS, Inoki CK, Shieh SR, Duffy TS, Kub FJ, Sturm JC (2002) Strain relaxation of SiGe islands on compliant oxide. J Appl Phys 91:9716–9722
67. Freund LB, Suresh S (2003) Thin film materials: stress, defect formation and surface evolution. Cambridge University Press, Cambridge
68. Huang R, Yin H, Liang J, Hobart KD, Sturm JC, Suo Z (2001) Relaxation of a strained elastic film on a viscous layer. MRS Symp Proc 695:115–120
69. Mei H, Huang R, Chung JY, Stafford CM, Yu HH (2007) Buckling modes of elastic thin films on elastic substrates. Appl Phys Lett 90:151902
70. Vella D, Bico J, Boudaoud A, Roman B, Reis PM (2009) The macroscopic delamination of thin films from elastic substrates. Proc Natl Acad Sci 106:10901–10906
71. Vandeparre H, Damman P (2008) Wrinkling of stimuloresponsive surfaces: mechanical instability coupled to diffusion. Phys Rev Lett 101:124301
72. Chung JY, Nolte AJ, Stafford CM (2009) Diffusion-controlled, self-organized growth of symmetric wrinkling patterns. Adv Mater 21:1358–1362
73. Chen X, Yin J (2010) Buckling patterns of thin films on curved compliant substrates with applications to morphogenesis and three-dimensional micro-fabrication. Soft Matter 6:5667–5680
74. Li B, Jia F, Cao Y-P, Feng X-Q, Gao H (2011) Surface wrinkling patterns on a core-shell soft sphere. Phys Rev Lett 106:234301
75. Khang D-Y, Rogers JA, Lee HH (2008) Mechanical buckling: mechanics, metrology, and stretchable electronics. Adv Funct Mater 18:1–11
76. Chung JT, Nolte AJ, Stafford CM (2011) Surface wrinkling: a versatile platform for measuring thin-film properties. Adv Mater 23:349–368
77. Yang S, Khare K, Lin P-C (2010) Harnessing surface wrinkle patterns in soft matter. Adv Funct Mater 20:2550–2564
78. Mei Y, Kiravittaya S, Harazim S, Schmidt OG (2010) Principles and applications of micro and nanoscale wrinkles. Mat Sci Eng R 70:209–224

Chapter 6
Crease Instability on the Surface of a Solid

Wei Hong and Fangliang Gao

Abstract Crease instability takes place on the surface of a solid when the lateral compression exceeds a critical level. Despite the weak singularity at the cusp, formation of a crease relieves the elastic energy through the local out-of-plane deformation. Due to the singular fields of deformation and stress, a crease differs from wrinkles in both appearance and mathematical description. The singularity also prevents common linearization-based methods for stability analysis from correctly predicting the critical condition for the instability. This chapter first reviews the recently published energy method which has successfully calculated the critical strain for crease formation. Examples of more general loading conditions, such as the crease instability on growing tissues, are demonstrated. Moreover, the equilibrium states of a crease on a film of finite thickness are studied. The approach of extrapolating the results of thinner films actually yields a more accurate result than the energy method. With the aid of this new approach, the later part of the chapter focuses on the effect of material properties on crease instabilities. The effect of material compressibility, as well as the crease formation on a swelling gel, is further investigated.

6.1 Introduction

When a block of rubber is compressed, creases may form laterally on the free surface. Just like wrinkles, creases relieve elastic energy by developing inhomogeneous out-of-plane deformation patterns. Creases differ from wrinkles in appearance as well as the detailed mode of deformation, even though the two terms are

---

W. Hong (✉) • F. Gao
Department of Aerospace Engineering, Iowa State University, Ames, IA 50011, USA
e-mail: whong@iastate.edu

X. Chen (ed.), *Mechanical Self-Assembly: Science and Applications*,
DOI 10.1007/978-1-4614-4562-3_6,

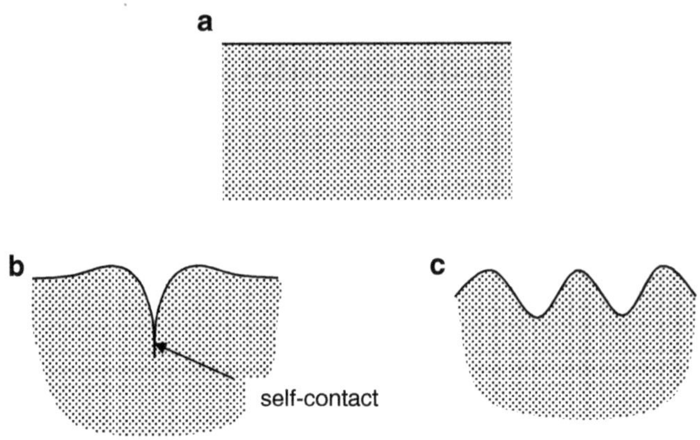

Fig. 6.1 Sketches of a crease and wrinkles on a free solid surface. (**a**) Homogeneous deformation of a solid being compressed horizontally. (**b**) A crease forms on the surface of the solid. (**c**) Wrinkles develop on the surface

used interchangeably in daily life. In this chapter, we define crease as the type of surface instability exhibiting an area of self-contact, as sketched in Fig. 6.1, and refer to the waviness of smaller amplitude as wrinkles. Because of the self-contact, creases only appear on a free surface, while wrinkles may occur on the interface of two solid materials. At a later stage, however, fully developed wrinkles with large amplitude may fold into each other and form creases on a free surface.

Creases are not only observed on the surface of soft materials under compression or bending (e.g., [1]) but also on swelling polymers [2, 3] and growing biological tissues [4, 5]. Creases are also seen on inelastic materials. Examples include the surface folding during metal forming and flow of highly viscous fluid, such as lava. Creases are often undesirable in applications: creases on a garden hose may induce fatigue failure; creases on fabric require extra treatment every time; creases on a coated surface may cause fracture or debonding of the coating. On the other hand, controlled surface creases are found to be useful in autonomous microscopic patterning (e.g., [6]). The mechanics of crease has attracted great interest recently [5, 7–13].

The tip of a mathematically sharp crease, with the radius of curvature being 0, naturally reminds one of a crack tip. Indeed, the deformation is singular at the tip. The geometry is similar to that of a 180° disclination—the initially flat surface folds back into itself around a line. Consider a 180° disclination formed by an incompressible material. Under plane-strain condition, the strain near the disclination center can be obtained directly from the geometry: the hoop and radial stretches are, respectively, $\lambda_\theta = \sqrt{2}$ and $\lambda_r = 1/\sqrt{2}$. The equilibrium, on the other hand, requires the stress to be singular near a disclination core, $\sigma \sim R^{-1/2}$, with R being the distance away from the core in the undeformed state. Similarly, we expect the strain to be finite (or at most having a removable singularity), while the stress is singular near the crease tip on an incompressible material. Numerical calculations using

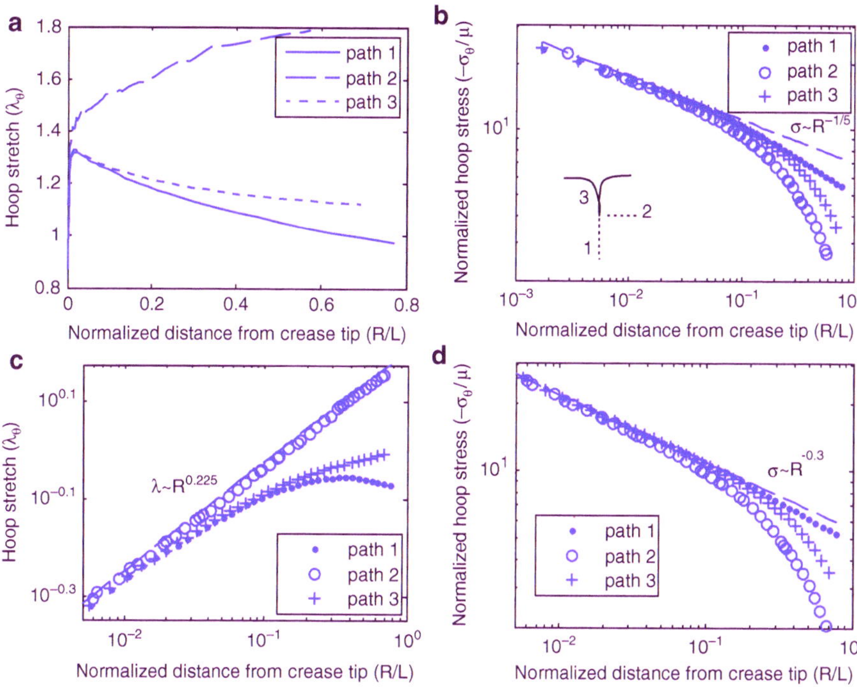

Fig. 6.2 Singularity of stresses and strains near a crease tip. Numerical results presented are the stresses and strains along three paths, as indicated by the *inset* of (**b**). (**a**) The strain is finite near a crease on an incompressible solid, except for the crease tip, a removable singularity. (**b**) Singularity of the hoop stress in an incompressible solid. (**c**) Stretch and (**d**) stress exhibit higher order singularity near a crease on a compressible material. All stresses are normalized by the initial shear modulus μ, and lengths are normalized by the crease depth L

finite-element method have confirmed our expectation, as plotted in Fig. 6.2a, b. The stress near a crease tip has a lower order of singularity than that of a disclination: approximately, $\sigma \sim R^{-1/5}$. The order of singularity is dependent on the compressibility of the material. As shown in Fig. 6.2c, d, both stress and strain have higher order of singularity for a compressible material. The stretches approximately scale with the distance from the crease tip as $\lambda \sim R^{0.225}$. In terms of strain, the singularity is logarithmic, $\varepsilon \sim \ln R$. The material law used to generate the numerical results in Fig. 6.2c, d is neo-Hookean with the ratio between shear and bulk moduli being $\alpha = 0.1$.

The geometric nonlinearity, including finite rotation and self-contact, and the singularity together have made the crease problem mathematically difficult to tackle. Recent studies have gained some understanding on the physics of the crease instability, while some fundamental questions still remain to be answered. This chapter summarizes some analytical and numerical approaches for crease analysis, and introduces a few applications of the theory.

6.2 Mathematical Description and Linear Stability Analysis

Let us take the undeformed state to be the reference and label all material particles by their reference coordinates X_K. The semi-infinite block under consideration has a free surface at $X_2 = 0$. We will first consider plane-strain deformation so that the motion of the material particles is limited in the $X_1 - X_2$ plane. The field of deformation is fully described by the mappings between the reference and the current coordinates, $x_i(X_1, X_2)$. In the absence of tearing or fracture, the mappings are continuous functions, and in most places differentiable. The deformation gradient tensor,

$$F_{iK} = \frac{\partial x_i}{\partial X_K} \tag{6.1}$$

is a measure of strain, and is related to the stress at the same material point by specific material laws. For example, the nominal stress of an incompressible neo-Hookean solid is related to the deformation gradient as

$$s_{iK} = \mu F_{iK} - pF_{Ki}^{-1}, \tag{6.2}$$

where F_{Ki}^{-1} is the inverse of the deformation gradient tensor, and p is an undetermined pressure field due to incompressibility. The incompressibility also poses a constraint on the deformation gradient:

$$\det F = 1. \tag{6.3}$$

The nominal stress field will need to satisfy the equilibrium equation

$$\frac{\partial s_{iK}}{\partial X_K} = 0, \tag{6.4}$$

and the traction-free boundary condition on the free surface:

$$s_{i2}(X_1, 0) = 0. \tag{6.5}$$

Finally, the remote deformation is assumed to be the prescribed uniform compression

$$[F_{iK}(\infty)] = \begin{bmatrix} \lambda & 0 \\ 0 & 1/\lambda \end{bmatrix}, \tag{6.6}$$

and the displacement needs to be properly constrained against rigid-body motion, e.g., on the symmetry plane, $x_1(0, X_2) = 0$ and $x_2(0, 0) = 0$. Equations (6.1) through (6.4), together with the boundary conditions, consist a boundary-value problem of the unknown fields $x_i(X_1, X_2)$ and $p(X_1, X_2)$. The uniform compression of the block is a trivial solution to the problem:

$$\begin{cases} \bar{x}_1 = \lambda X_1, \\ \bar{x}_2 = \frac{X_2}{\lambda}, \\ \bar{p} = \frac{\mu}{\lambda^2}. \end{cases} \tag{6.7}$$

The nontrivial question for crease and wrinkle instabilities is whether there exists a nonuniform solution to the boundary value problem, and if so, at which critical stretch λ_c.

Biot may be the first to have analyzed the wrinkle instability on the surface of a semi-infinite viscoelastic solid [14] and that on an incompressible neo-Hookean solid [15] by using a linear stability analysis. The method he used is essentially the same as the following derivation. Let us introduce a small perturbation on the trivial solution

$$\begin{cases} x_1 = \bar{x}_1 + \sin k\lambda X_1 \left[A \exp(kX_2/\lambda) + B\lambda^2 \exp(k\lambda X_2)\right] \\ x_2 = \bar{x}_2 - \cos k\lambda X_1 \left[A \exp(kX_2/\lambda) + B \exp(k\lambda X_2)\right] \\ p = \bar{p} + kA\left(\lambda^2 - \lambda^{-2}\right) \cos k\lambda X_1 \exp(kX_2/\lambda) \end{cases}, \tag{6.8}$$

and linearize (6.2) and (6.4) by keeping only terms linear in kA and kB. It can be verified that the specific functional form in (6.8) satisfies the equilibrium equation and the incompressibility constraint to the first order of kA and kB. The traction-free boundary condition (6.5) becomes a linear eigenvalue problem for A and B. The condition for the existence of nontrivial solutions is then given by

$$\lambda^3 + \lambda^2 + \lambda - 1 = 0. \tag{6.9}$$

The critical stretch for the instability is the positive real root of (6.9), approximately,

$$\lambda_w \approx 0.5437, \tag{6.10}$$

which corresponds to a compressive strain of $\varepsilon_w \approx 46\%$. This critical value, however, exceeds the experimentally measured value, $\varepsilon_c \approx 35\%$ [1, 8]. As indicated by the deformation mode of instability in (6.8), ε_w is the critical strain for wrinkle instability instead of crease instability.

Reexamining the procedure of the linear stability analysis, one may find that the method is flawed for analyzing the crease instability. The error occurs at the linearization step. It is true for arbitrarily small crease, the continuous displacement may always be written into Fourier series of X_1, as in (6.8). However, its derivatives are not necessarily small. In fact, the strain near a crease tip will have removable or logarithmic singularity, independent of the crease amplitude. Therefore, linearizing the equations by keeping only the terms linear in kA and kB is an undependable approach. Nevertheless, its weak singularity may allow the actual strain field to be approximated by a truncated series containing higher order terms in kA and kB. In their nonlinear post-buckling analysis [13], Cao and Hutchinson find the instability modes to be more crease-like as more terms are retained. For the onset of the crease instability, such a practice will result in a nonlinear eigenvalue problem (i.e., to identify the condition for the existence of nontrivial solutions to a set of nonlinear algebraic equations) which does not have a general method of solving.

6.3 Energetic Approach

In Sect. 2, it is demonstrated that Biot's linear approach is unsuitable for analyzing the crease instability. On the other hand, nonlinear analysis may face the difficulty of solving a nonlinear eigenvalue problem. Despite the lack of a closed-form analytical solution, the crease instability can still be studied numerically. In fact, the formation of a crease can be directly simulated by any finite-element software capable of handling nonlinear geometry and contact. For direct simulation, an initial defect is needed and the defect amplitude needs to be large enough compared to the size (or wavelength) of the defect [13], in order to reveal the instability at a critical strain close to 35%. In this section, we review a numerical method which analyzes the problem from an energetic approach [10, 11].

A major shortcoming of the linear stability analysis is the inability to capture the singular geometry of the creased surface, and thus the actual instability mode is not among the candidates of the linearized eigenvalue problem. On the other hand, the characteristic features of a crease are already known: a region of self-contact, and a 180° folding. In numerical calculation, one can prescribe the features by specifying a displacement boundary condition. Half of the creased body is modeled as shown in Fig. 6.3. A horizontal displacement equals the horizontal coordinate of each point is prescribed on a section of the surface, OA, to deviate the block from the homogeneous deformation state. Such a crease may require traction on the inseam to maintain equilibrium. Instead of studying the real crease problem with unknown contact area, we will compare the difference in elastic energy between the uniform deformation state and the creased state of prescribed crease depth. The depth of the crease (measured in the reference state) L is taken to be much smaller than the size of the block so that L is the only length scale in the problem. The elastic energy per unit thickness of the creased body minus that of the homogeneously deformed body is, by dimensional consideration,

$$\Delta U = \mu L^2 f(\varepsilon), \tag{6.11}$$

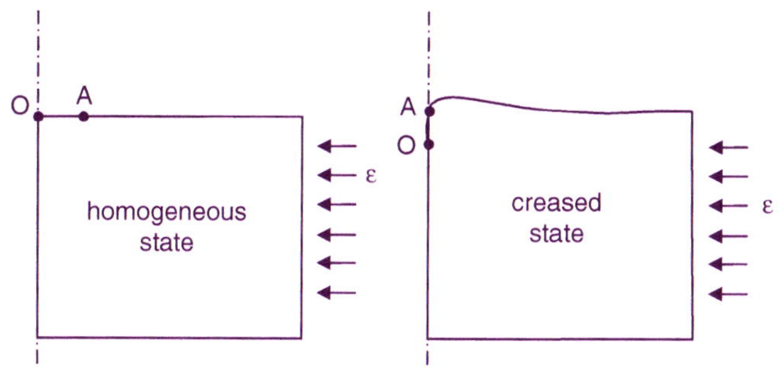

Fig. 6.3 Sketch of the computational domain. Only half of the creased body is modeled. A displacement is prescribed on OA at the surface of the block under plane-strain compression, to form the crease

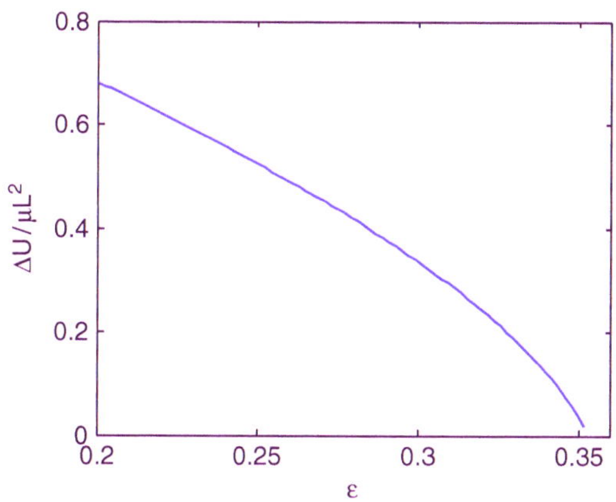

Fig. 6.4 The elastic energy difference between the creased state and the homogeneous state, plotted as a function of the applied compressive strain

where $f(\varepsilon)$ is a dimensionless function of the nominal strain ε. $f(\varepsilon)$ can be numerically computed using most finite element program. Figure 6.4 shows the result obtained by using SIMULIA Abaqus.

Under no compression, the homogeneous state is stress-free while the creased state is stressed, so $f(0)>0$. When the block is being compressed, the crease relieves the locally compressive stress, and $f(\varepsilon)$ is a decreasing function of ε. At a critical strain $\varepsilon_c \approx 35\%$, the energy of a creased state equals that of the homogeneous state, $f(\varepsilon_c) = 0$. In other words, the homogeneous deformation of the block becomes a metastable state. Beyond the critical point, no state of equilibrium exists for the prescribed crease depth and the computation of $f(\varepsilon)$ cannot be continued. The semi-infinite block lacks a length scale. Following (6.11), when $\varepsilon>\varepsilon_c$ and $f(\varepsilon)$ <0, the elastic energy of the creased state of any crease depth L is lower than that of the homogeneous state, and the deeper the crease, the lower the elastic energy.

Therefore for a semi-infinite block, a crease will extend indefinitely without reaching equilibrium. For a finite block, on the other hand, the crease will equilibrate at a finite depth, as discussed in Sect. 5.

Such a numerical approach has clearly shown the critical strain for the crease instability, ε_c, beyond which the creased state is energetically favorable over the homogeneous state. Although an energy barrier may be present between the two states, such a result at least shows that the crease will form at the critical strain under large enough perturbation.

6.4 General Deformation States

Next let us consider the crease instability under general loading conditions. Consider a semi-infinite elastomer block under homogeneous deformation, with λ_1 and λ_3 being the stretches parallel to the free-surface, and λ_2 the stretch perpendicular to the surface. The compression in the x_1-direction is taken to be more severe than that in the x_3-direction, so that a crease forms with the ridge parallel to the x_3-direction, leaving the stretch λ_3 unchanged. The crease is in a generalized plane-strain state. To analyze the instability of the crease, we imagine an intermediate state to bridge between the referenced stress-free state and the state of actual deformation. Figure 6.5 illustrates the deformation in the three states. A unit cube in the reference state deforms with principal stretches λ_1, λ_2, and λ_3 in the current state. The intermediate state retains the same stretch in the x_3-direction, λ_3, but is stress-free in the other two directions. For an incompressible material, the stretches of the intermediate state are $\lambda_1^0 = \lambda_2^0 = 1/\sqrt{\lambda_3}$ and $\lambda_3^0 = \lambda_3$. Taking the intermediate state as the

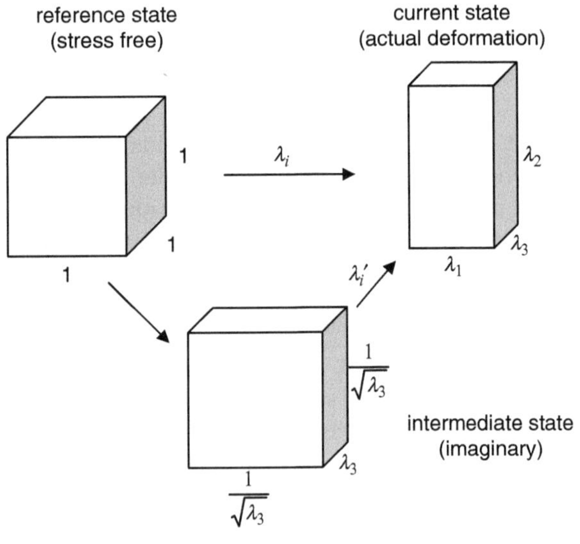

Fig. 6.5 The deformation of a generalized plane-strain problem, and the equivalent plane-strain problem with the aid of an intermediate state

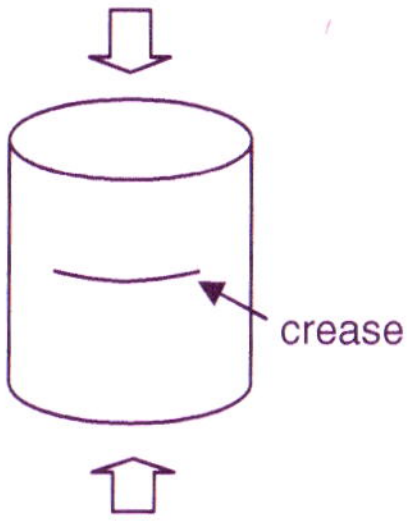

Fig. 6.6 The critical compressive strain for creasing on the surface of an axially compressed cylinder is $\varepsilon = 44\%$

new reference, we can regard the relative deformation of the current state as plane-strain. The relative stretches with respect to the new reference are $\lambda'_1 = \lambda_1\sqrt{\lambda_3}$, $\lambda'_2 = \lambda_2\sqrt{\lambda_3}$, and $\lambda'_3 = 1$. The elastic energy per unit volume of the material can also be written in terms of the relative stretches. For a neo-Hookean material,

$$W(\lambda'_1, \lambda'_2) = \frac{\mu}{2}\left(\lambda_1^2 + \lambda_2^2 + \lambda_3^2 - 3\right) = \frac{\mu}{2\lambda_3}\left(\lambda_1'^2 + \lambda_2'^2 + \lambda_3^3 - 3\right). \tag{6.12}$$

The strain-energy function would give rise to the same stress–strain relation as that of a neo-Hookean material under plane-strain deformation. The effective shear modulus of the generalized plane-strain problem is μ/λ_3.

Such a similarity provides a means of utilizing the result obtained in Sect. 3 to the crease instability of a material under general deformation state. The critical strain $\varepsilon_c \approx 35\%$ corresponds to a critical stretch of $\lambda_c \approx 0.65$ in a plane-strain state. The same critical condition is applicable to a generalized problem in which $\lambda'_c \approx 0.65$. In terms of the stretch with respect to the stress-free reference, the critical condition is $\lambda_1\sqrt{\lambda_3} \approx 0.65$. Since the material is assumed to be incompressible, $\lambda_1\lambda_2\lambda_3 = 1$, the critical condition for crease instability can be rewritten as

$$\frac{\lambda_2}{\lambda_1} \approx 2.4. \tag{6.13}$$

Here, λ_1 is the stretch in the direction perpendicular to the crease line and parallel to the free surface, while λ_2 is the stretch in the normal direction of the free surface. It is noteworthy that the same procedure is also applicable to the linear stability analysis for wrinkles, and it will translate the critical strain for wrinkle instability $\varepsilon_w \approx 46\%$ to a generalized plane-strain state as $\lambda_2/\lambda_1 \approx 3.4$.

As an example, consider an elastomer cylinder being compressed axially. Crease may form circumferentially on the cylinder surface, as sketched in Fig. 6.6. To predict the onset of the crease, we may use the critical condition (6.13) directly. Let λ_z be the axial stretch. The corresponding radial and circumferential stretches are $\lambda_r = \lambda_\theta = 1/\sqrt{\lambda_z}$. The critical condition for crease instability is $\lambda_r/\lambda_z \approx 2.4$, or $\lambda_z = 0.56$. In terms of compressive strain, crease forms at approximately 44% of axial compression. (The linear perturbation result will instead predict a critical strain of 56%.) For an infinitesimal crease with no intrinsic length scale, the critical condition will not be affected by the curvature of the surface.

The critical condition for a general deformation state has also been used to study the creases generated by the growth of soft tissues [5]. We will illustrate this application by looking at two simple examples regarding the tissue growth on a concave surface and a convex surface. For simplicity, we will assume the growth to be isotropic and homogeneous, with the growth tensor $\mathbf{G} = g\mathbf{I}$, where $\mathbf{I}$ is the identity tensor [16]. Such a growth tensor corresponds to the state of growth in which the tissue will expand isotropically with linear stretch g if no constraint is posed. First, let us look at the growth on the outer surface of a rigid spherical core of radius R. Denote the thickness of the tissue layer before growth as H. Assuming the tissue to be incompressible, one may obtain the field of deformation without solving the equilibrium equations. The radius of the external surface r after growth is related to g as

$$r^3 - R^3 = g^3\left[(R+H)^3 - R^3\right]. \tag{6.14}$$

The circumferential stretch on the external surface of the tissue, relative to a stress-free isotropic state of growth, is

$$\lambda_\theta = \left[1 - \left(1 - \frac{1}{g^3}\right)\left(\frac{R}{R+H}\right)^3\right]^{1/3}. \tag{6.15}$$

The critical linear growth ratio for crease instability, g_c, can then be obtained by applying the critical condition for equal biaxial compression, $\lambda_\theta^{\frac{3}{2}} \approx 0.65$:

$$g_c \approx \left[1 - 0.58\left(1 + \frac{H}{R}\right)^3\right]^{-1/3}. \tag{6.16}$$

In the case of a relatively thin tissue, $H << R$, $g_c \approx 1.33 + 1.8H/R$.

The similar procedure can be applied to the growth of a tissue on the inner surface of a rigid shell. Denoting the radius of the rigid wall as R and the thickness of the tissue before growth as H, one can easily obtain the critical linear growth for crease instability:

$$g_c \approx \left[1 - 0.58\left(1 - \frac{H}{R}\right)^3\right]^{-1/3} \approx 1.33 - 1.8H/R. \tag{6.17}$$

Obviously, the tissue growth on a concave surface would induce a crease more easily than that on a convex surface. In the limiting case of a flat surface, both results reduce to the critical condition of approximately 33% of isotropic growth for crease instability to take place.

6.5 Crease on a Film of Finite Thickness

The failure of the linear stability analysis in Sect. 2 suggests that a perturbation in the form of a singular function may be needed for the crease instability. However, the introduction of a non-smooth or a singular perturbation would immediately give rise to geometric nonlinearity which prohibits common analytical methods. The simplest form of a singular perturbation may be adding a concentrated load on the surface of an elastomer block under lateral compression. As sketched in Fig. 6.7a, a force per unit thickness of the block, F, is applied perpendicular to the free surface, and the loading point displaces by δ relative to the unperturbed surface. It is expected that the onset of the crease instability is the point at which the corresponding stiffness of the point load vanishes, i.e., when a point on the surface can displace spontaneously in the absence of any force. The corresponding problem in linear elasticity is known as the Flamant problem [17], and has a closed-form solution. The large deformation, however, would always introduce geometric nonlinearity as well as self-contact near the singular point. Research on the analytical solution of the finite-deformation Flamant problem and Boussinesq problem has been carried out (e.g., [18, 19]). However no analytical result for the compressive force case has been found up till the writing of this chapter. In fact, a crease will always form at the loading line, regardless of the magnitude of the load. Here in this section, we take the numerical approach by using the finite element method. It is noteworthy that although the analytical solution for the displacement at tip is infinity, the numerical solution is finite because the load is spread over one element through the course of discretization.

Some representative load–displacement curves are shown in Fig. 6.7b. Although a semi-infinite block is preferred for theoretical analysis, the numerical model has a finite size. The displacement is normalized by the overall thickness of the block, H,

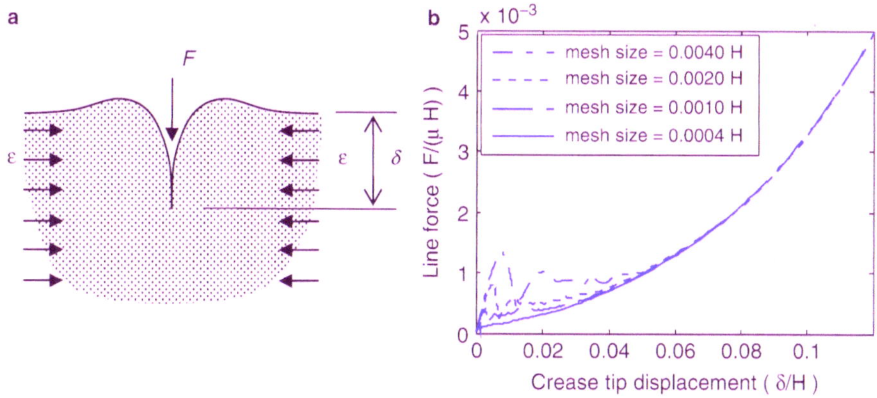

Fig. 6.7 (**a**) Sketch of the finite-deformation Flamant problem: a line force acting on a semi-infinite block under latter compression. (**b**) The load–displacement curves and the mesh dependence

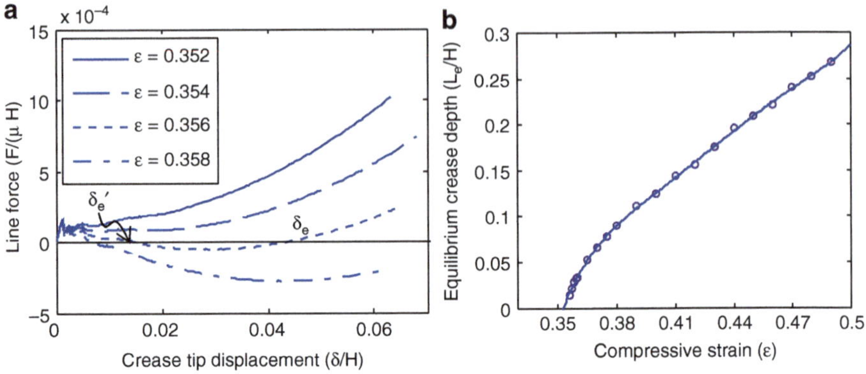

Fig. 6.8 (**a**) The load–displacement curve of a line force on the surface of a large block of size H, under various levels of compressive strain. The two intersections with the horizontal axis correspond to two equilibrium states, one being unstable and the other stable. (**b**) The depth of a crease in stable equilibrium, plotted as a function of the lateral compressive strain. The *circles* are the numerical results, and the *line* is a guide to the eye

and the line force by μH. At the beginning of the curve, when the displacement is smaller than one element, the small strain approximation holds in most elements, so the numerical results recover the linear solution. However, such a solution is never physical, since it is highly dependent on the mesh size, as shown in Fig. 6.7b. As soon as contact takes place in one element, the force–displacement curve deviates from the linear solution. At a displacement about one order of magnitude larger than the mesh size, the numerical result becomes more stable and mesh independent.

Disregarding the nonphysical linear solution at displacements smaller than an element, the slope of the load–displacement curves does show a dependence on the lateral compressive strain. As shown in Fig. 6.8a, the sign of the initial slope changes at a critical compressive strain $\varepsilon_c \approx 35\%$. However, due to the artifact at small displacements, the accurate values of the initial slopes are difficult to obtain. Here we take an alternative approach by looking at the horizontal intercept of the load–displacement curve. Under a supercritical compressive strain, $\varepsilon > \varepsilon_c$, the load–displacement curve intersects with the horizontal axis at two points, δ'_e and δ_e, as shown in Fig. 6.8a. These two intersections correspond to two equilibrium states, in which no force is needed and the crease forms spontaneously. The intersection δ'_e is unstable and is induced by the numerical artifact at small displacement. An infinitely accurate solution is expected to have negative initial slope and intersect with the horizontal axis at only one point. Consider only the physical equilibrium point δ_e. Figure 6.8b plots the corresponding crease depth as a function of the applied strain, $L_e(\varepsilon)$. In contrast to the prescribed crease under a subcritical strain as described in Sect. 3, here the crease forms spontaneously in the absence of any external force or constraint ($F = 0$). At a depth L_e, the creased state is in equilibrium and energetically stable.

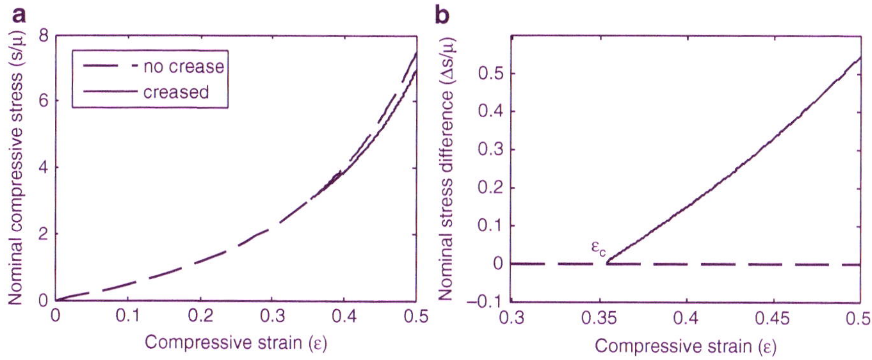

Fig. 6.9 (**a**) The remote nominal compressive stresses as functions of the overall compressive strain, for the homogeneous compression state and the creased state. The crease relieves the compressive strain. (**b**) The difference in remote stress between the creased state and the homogeneous state, plotted as a function of the overall compressive strain

This calculation, in turn, reveals the equilibrium crease depth on a film of thickness H bonded to a rigid substrate. With H being the only length scale of the problem, the equilibrium crease depth L_e scales with H and is a monotonically increasing function of the compressive strain at $\varepsilon > \varepsilon_c$. Extrapolating the curve of equilibrium crease depths, Fig. 6.8b, an intersection with the horizontal axis could be found. This intersection is the strain at which an infinitesimal crease is at a stable equilibrium state, i.e., the onset of the crease instability ε_c. The numerical value $\varepsilon_c \approx 35\%$ confirms the result obtained in Sect. 3.

Lack of an intrinsic length scale, the behavior of a film of finite thickness, is indifferent from that of a semi-infinite block. At the critical strain, an infinitesimal crease nucleates. The contact depth of the crease then increases with the applied strain. In terms of displacement, the state of an infinitesimal crease is infinitely close to a homogeneous deformation state. In other words, perturbation on the displacement field with infinitesimal amplitude (but finite in derivative) would induce the energetically more stable crease. In atomic scale, the discontinuous thermal fluctuation serves well as such a perturbation. On the other hand, the surface of a solid, especially polymer, is neither perfectly flat nor atomically smooth. We may study the energy landscape near the critical point by looking at the work done by the remote stress. Figure 6.9 plots the remote stresses as functions of the overall compressive strain, for both the homogeneous state and the creased state. In a reversible process, the work done by the remote stress (i.e., the area below the curve multiplied by the volume of the bulk) equals the elastic energy stored in the bulk. Beyond the critical strain ε_c, the creased state equilibrates with a lower remote stress, and thus stores less energy. At the critical strain, the two states are in equilibrium with the same remote stress, and bare the same elastic energy. In order to obtain the results for the creased state, creases with different depths are prescribed at various levels of the applied strain. If a crease initiates at infinitesimal depth before reaching equilibrium at a finite depth, the calculation shows that there is no energy barrier between the homogeneous state and the creased state. Nevertheless, in the actual physical process, a finite energy barrier may be present between the homogeneously deformed state and a creased state. Such an energy barrier may be due to various reasons, including the effect of surface or the material heterogeneity near a surface, the effect of inelastic deformation near a crease tip, etc. The existence of a finite energy barrier may give rise to a length scale. (In fact, the heterogeneity near the surface of a material may define a length scale directly.) The intrinsic length scale may delay the crease instability for that a crease shallower than a specific depth may be energetically unfavorable. For an incompressible neo-Hookean material, such an effect is minor. Based on the numerical results shown in Fig. 6.8b, an energy barrier which prevents the formation of a crease shallower than 5% of the overall thickness of a bonded film would only increase the critical strain by 1%. The effect of energy barrier could be more significant to other materials, such as a swollen polymeric gel, as is discussed later in Sect. 6.

6.6 Effect of Material Properties

Till now, we have only been looking at the crease instability on an incompressible neo-Hookean material. The agreement between the calculated critical strain $\varepsilon_c \approx 35\ \%$ and experimental results may be attributed to the finite but moderate strain near the crease tip. Indeed, for an incompressible neo-Hookean material, the stretch near the tip is close to that of a 180° disclination, $\lambda_\theta = \sqrt{2}$, as discussed in Sect. 1. Under such a moderate stretch, neo-Hookean model is a good approximate for many hyperelastic materials. However, for a compressible material, the radial distributions of both stress and strain are singular near the crease tip. Therefore, it is expected that the critical condition for crease instability is dependent on the detailed constitutive relations as well as the compressibility of materials.

We will first look at the effect of compressibility on the critical strain for crease. As of yet, we have only numerical methods to determine the onset of crease instability, e.g., through extrapolating the curve in Figs. 6.4, 6.8b, or 6.9b. Using the commercial finite-element code SIMULIA Abaqus 6.10, we calculate the data points in Fig. 6.10a using a set of curves as the one in Fig. 6.9b. The constitutive relation used is the compressible neo-Hookean material law, with the elastic energy density being

$$W = \frac{\mu}{2}\left(\bar{I}_1 - 3\right) + \frac{K}{2}(J-1)^2, \tag{6.18}$$

where $\bar{I}_1 = J^{-\frac{2}{3}} F_{iM} F_{iM}$ is the first invariant of deviatoric deformation tensor, $J = \det F$ is the volume ratio of deformation, and K is the bulk modulus. We assume plane-strain deformation, and measure the compressibility in terms of the

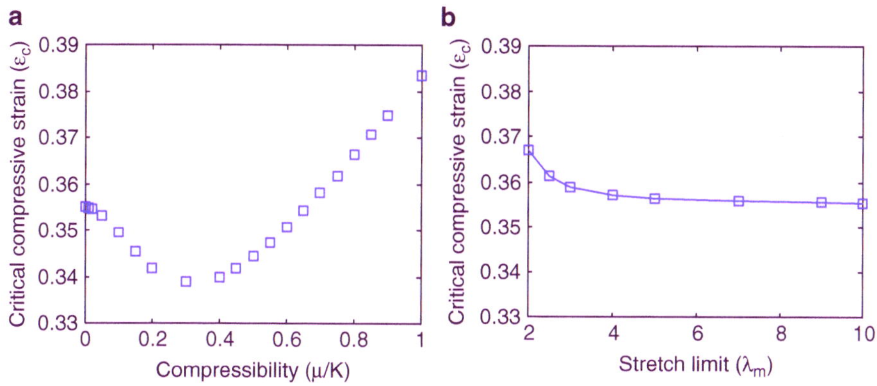

Fig. 6.10 (**a**) The dependence of the onset of crease instability on the compressibility of a neo-Hookean material. (**b**) The dependence of the crease instability on the extension limit of an incompressible Arruda–Boyce material

ratio between the two moduli, μ/K. Despite the difficulty in error control due to the nature of the finite element method and the singularity of the problem, the results plotted in Fig. 6.10a clearly show the dependence of the critical strain on the compressibility of the material. The critical strain for the crease instability, ε_c, is not a monotonic function of the compressibility. For a slightly compressible material, $\mu/K \leq 0.3$, the critical strain decreases with the increase in compressibility; for a highly compressible material, $\mu/K \geq 0.4$, the critical strain increases with compressibility. That is, crease forms more easily on a slightly compressible material than on an incompressible material, but more difficultly on highly compressible foam. Such dependence may be attributed to the two competing effects of the lateral constraint. Near the crease tip, the lateral constraint tends to delay the formation of the crease because of the local large deformation (and thus high constraint force). Remote to the crease tip, the lateral constraint increases the elastic strain energy at the same compression, and thus favors the formation of a crease. The numerical results show that a neo-Hookean material with intermediate compressibility, $\mu/K \approx 0.3$, is most susceptible to crease instability.

Now let us turn to the effect of the material constitutive behavior. Since only numerical methods are available, it is impossible to enumerate all material laws and test them one by one. On the other hand, the major difference among various constitutive relations of elastomers is at the high-extension response. Due to the extension limit of polymer chains, most elastomers have strain-stiffening effect. To study the effect of strain stiffening on crease instability, we invoke the commonly used Arruda–Boyce model with a free-energy function dependent on the stretch limit parameter λ_m [20]. Plotted in Fig. 6.10b is the critical strain for crease instability on a semi-infinite block of the Arruda–Boyce material, as a function of the stretch limit λ_m. It can be seen that the strain-stiffening effect tends to delay the onset of crease instability, for the simple reason that the distorted core of a crease stores higher elastic energy in a strain-stiffened material than that in a neo-Hookean material. However, because of the weak singularity and the moderate deformation around a crease tip as shown in Sect. 1, the effect is minor and may be of importance only when the stretch limit is very low. Figure 6.10b only plotted the results for an incompressible material while the behavior of compressible materials is very similar.

The crease instability on the surface of a swollen polymeric gel was observed in 1960s [Southern] and has recently attracted more interest [3, 7, 9, 12, 21]. Here as an example, we would use the numerical methods described in previous sections to study the crease instability on a polymeric gel. Due to the presence of the mobile solvent molecules in a swollen gel, its deformation is usually time dependent. Therefore, the critical condition for crease instability is also expected to be dependent on the deformation rate. We will consider two limiting cases.

Under very high deformation rate, the solvent molecules will not have time to migrate, and a gel appears to be incompressible. It was commented by Hong et al. [10, 11] that the critical condition for a crease that forms instantaneously can be derived directly from the critical strain of an incompressible elastomer. If a swollen gel is compressed instantaneously in a plane-strain state, the critical strain with respect to the free-swelling state is $\varepsilon_c \approx 35\,\%$. Similarly, during the fast swelling process of a gel bonded to a rigid surface, a crease forms instantaneously at a linear swelling ratio of $\lambda_c' \approx 2.4$, relative to the as-bonded state. The result is obtained by imagining that the bonded swollen state is achieved through instantaneously compressing an isotropic swollen gel of the same volume to match the lateral constraint from the rigid substrate.

At the slow deformation limit, the migration of solvent molecules has completed and a thermodynamic equilibrium can thus be assumed. To study the crease instability at this limit, we utilize the free-energy function for polymeric gels proposed by Hong et al. [22]:

$$W = \frac{\mu}{2}\left(\bar{I}_1 - 3 - 2\ln J\right) - \frac{kT}{v}\left[(J-1)\ln\frac{J}{J-1} + \frac{\chi}{J}\right] - \frac{\bar{\mu}}{v}(J-1), \tag{6.19}$$

in which kT is the temperature in the unit of energy, v the volume per solvent molecule, χ a dimensionless parameter for the enthalpy of mixing, and $\bar{\mu}$ the chemical potential of the solvent in the gel referenced to pure solvent. We implement the free-energy function into a user-defined subroutine in SIMULIA Abaqus 6.10 [10, 11], and take the following representative values of the material parameters: $v\mu/kT = 0.001$ and $\chi = 0.1$. A gel represented by this model increases its volume by approximately 39 times from a dry state to an isotropic swollen state in equilibrium with pure solvent, $\bar{\mu} = 0$. We first consider the procedure shown in Fig. 6.11a. A semi-infinite gel first swells by λ_0 in all directions, and is then subject to a lateral compression. The second step is an equilibrium process, in which the chemical potential of the solvent $\bar{\mu}$ is kept constant, and the deformation from the isotropic swelling state is assumed to be plane-strain. The critical strain is obtained

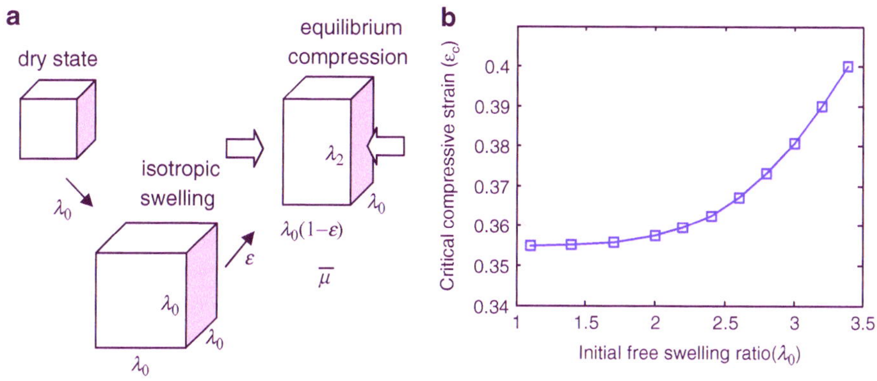

Fig. 6.11 (**a**) The two-step loading process of a block of swollen gel under compression. (**b**) The dependence of critical compressive strain for crease instability, relative to the isotropic swelling state, on the initial isotropic swelling ratio of the gel, λ_0

using the same method as shown in Fig. 6.9b. The numerical results are plotted in Fig. 6.11b. Due to the finite compressibility of a swollen gel, the critical compressive strain takes a value higher than that of an incompressible neo-Hookean rubber. The difference is more significant at higher initial swelling ratios, when the resulting gel tends to be more compressible.

The crease instability on a swollen gel being compressed mechanically is directly related to our earlier discussion, but may be of less interest practically. For polymeric gels that are responsive to various types of stimuli in the form of large deformation, the crease instability during the course of swelling is studied more often [7, 23]. Here, using a numerical method similar to that described in Fig. 6.9b, we also calculate the critical condition for the crease instability on a gel which undergoes constrained swelling. The loading condition is sketched in Fig. 6.12a. A large film of gel is bonded to a rigid substrate after initial isotropic swelling λ_0, and then swells more in the thickness direction, with the lateral directions constrained. Thermodynamic equilibrium is assumed for the whole deformation process, i.e., $\bar{\mu}$ is kept constant. The onset of crease instability is measured in terms of the relative swelling ratio λ'_c and plotted in Fig. 6.12b, as a function of the ratio of initial swelling at which the gel is bonded. The critical swelling ratio $\lambda'_c \approx 2.4$ for instantaneous crease formation is also plotted for comparison. It could be seen that at low initial swelling ratio, due to the limited compressibility, the critical condition for equilibrium crease is close to that of instantaneous instability, while at high initial swelling ratio, a crease forming in equilibrium requires a much lower relative swelling. With the combination of material parameters picked, if a gel is bonded (or cross-linked) at an initial swelling ratio of 2 or greater (~90% vol of solvent), an additional 200% swelling in the thickness direction would already induce crease instability in the long-term limit.

The plot in Fig. 6.12b may also serve as a dynamic phase diagram for the crease instability on a gel under constrained swelling. At a relative swelling smaller than

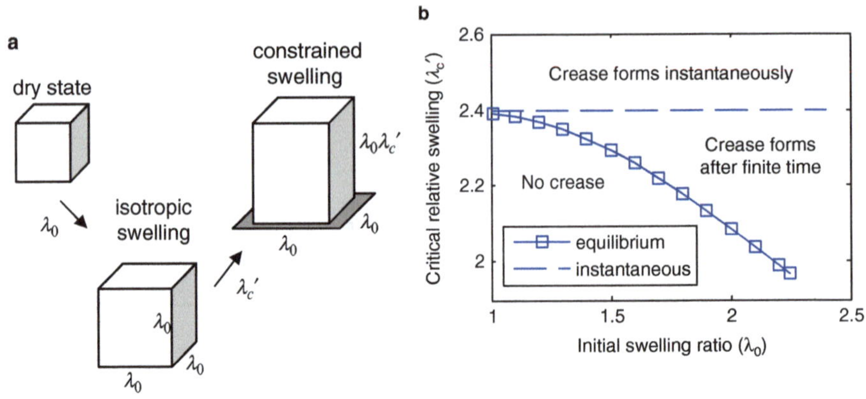

Fig. 6.12 (**a**) Sketch of the constrained swelling process. The gel is laterally constrained when bonded to a rigid substrate at initial swelling ratio λ_0. (**b**) The additional vertical swelling needed in an equilibrium state for crease to initiate at the free surface, λ_0', plotted as a function of λ_0. The dash line shows the critical swelling for instantaneous crease formation

the critical condition in equilibrium, crease never forms and a flat surface is stable. At a relative swelling ratio between the equilibrium condition and the instantaneous condition, a crease will form after a finite time, i.e., the time needed for the creased state to reach equilibrium. Such a phase diagram may be used to explain the diverse values (2–3.7) of critical swelling ratios measured in experiments [2, 9]. However, strictly speaking, the deformation rate of a system free of length scale would be difficult to measure. In fact, the time needed for solvent migration to equilibrate an infinitesimal crease is infinitesimally short. In other words, an infinitesimal crease takes no time to reach equilibrium. The situation would be different if the formation of a crease has a finite energy barrier or the material has an intrinsic length scale. Since a crease would then initiate at a finite size in order to be energetically favorable, and the migration of solvent molecules requires finite time to form the equilibrium crease, the apparent onset of the crease formation would be delayed. The amount of delay in terms of swelling ratio would thus be dependent on the speed of swelling as well as the thickness of the gel film. The heterogeneous distribution of solvent through the thickness of the gel during swelling may also play a role in determining the crease instability. To accurately predict the crease instability in a swelling gel, a dynamic model for crease instability is required.

6.7 Concluding Remarks

This chapter introduces a special type of instability, crease, which takes place on the free surface of a material being compressed. Characterized by the self-contact and the weak singularity in stress and strain, a crease has inherited geometric nonlinearity even at infinitesimal amplitude. The finite deformation prohibits the usage of the common linear stability analysis, which in turn predicts the wrinkle instability. The deformation mode of crease is not among the candidates of the linear stability analysis. While it may be probed through the introduction of a singular perturbation, or by considering higher order terms in stability analysis, the resulting nonlinear eigenvalue problem is not readily solvable.

Despite the mathematical difficulty in finding a closed-form solution, the onset of instability could be obtained numerically by extrapolating the solutions of a prescribed crease of finite size to those of an infinitesimal crease. For an incompressible neo-Hookean material under plane-strain deformation, the critical compressive strain for crease instability is $\varepsilon_c \approx 35\%$. This value is very close to the experimentally measured onset of crease on rubbery materials, and is significantly lower than the onset of wrinkle instability, $\varepsilon_w \approx 46\%$, as given by the linear stability analysis. For a more general deformation mode, the critical point can be obtained from this result by utilizing the specialty of the neo-Hookean material law. Under a generalized plane-strain compression with stretch λ_3 in the third direction, the critical condition is $\lambda_1\sqrt{\lambda_3} \approx 0.65$ or $\lambda_2/\lambda_1 \approx 2.4$, where λ_1 is the stretch along the direction of compression. The results can be applied directly to crease instability in other similar scenarios. During the growth of a tissue, the onset of crease on a laterally constrained tissue on a flat substrate is an isotropic growth rate of 1.33 in the absence of the constraint. A concave substrate would promote the crease instability while a convex substrate would delay the crease formation.

The dependency of the crease instability on the material constitutive relations is also studied using the numerical methods. It is found that the material compressibility has a strong influence on the crease instability, while the strain-stiffening at high stretching has very small effect. More interestingly, the dependence of the critical strain is not a monotonic function of the compressibility. A slightly compressible material favors the crease instability while highly compressible foam would postpone or prevent the crease formation. Polymeric gels, on the other hand, have time-dependent mechanical behaviors. The short-term response of polymeric gel is just like an incompressible rubber, while the long-term response which brings a gel to thermodynamic equilibrium is highly compressible because of the solvent migration. It is found in this chapter that the critical strain for long-term crease instability is always lower than that of an instantaneously formed crease. The limiting results in the two extremes predict the upper and lower bounds for crease instability, while the onset of crease on a polymeric gel would depend on the actual loading process as well as the initial condition.

Although there have been some advances on the understanding of crease instability, a few fundamental questions remain unanswered: What is the ultimate mathematical method for analyzing instabilities that has inevitable nonlinearity? What is the role of material heterogeneity or surface effect? What is the critical condition for a crease which forms dynamically?

References

1. Gent AN, Cho IS (1999) Surface instabilities in compressed or bent rubber blocks. Rubber Chem Technol 72:253
2. Southern E, Thomas AG (1965) Effect of constraints on the equilibrium swelling of rubber vulcanizates. J Polym Sci A 3:641
3. Tanaka T, Sun S-T, Hirokawa Y, Katayama S, Kucera J, Hirose Y, Amiya T (1987) Mechanical instability of gels at the phase transition. Nature 325:796
4. Hrousis CA, Wiggs BJR, Drazen JM, Parks DM, Kamm RD (2002) Mucosal folding in biologic vessels. J Biomech Eng 124:334
5. Jin L, Cai S, Suo Z (2011) Creases in soft tissues generated by growth. EPL 95:64002
6. Guvendiren M, Burdick JA, Yang S (2010) Solvent induced transition from wrinkles to creases in thin film gels with depth-wise crosslinking gradients. Soft Matter 6:5795–5801
7. Hwa T, Kardar M (1988) Evolution of surface patterns on swelling gels. Phys Rev Lett 61:106–109
8. Ghatak A, Das AL (2007) Kink instability of a highly deformable elastic cylinder. Phys Rev Lett 99:076101
9. Trujillo V, Kim J, Hayward RC (2008) Creasing instability of surface-attached hydrogels. Soft Matter 4:564
10. Hong W, Zhao X, Suo Z (2009) Formation of creases on the surfaces of elastomers and gels. Appl Phys Lett 95:111901
11. Hong W, Liu Z, Suo Z (2009) Inhomogeneous swelling of a gel in equilibrium with a solvent and mechanical load. Int J Solids Struct 46:3282–3289
12. Wong WH, Guo TF, Zhang YW (2010) Surface instability maps for soft materials. Soft Mater 6(22):5743–5750
13. Cao Y, Hutchinson JW (2011) From wrinkles to creases in elastomers: the instability and imperfection-sensitivity of wrinkling. Proc R Soc A 468(2137):94–115. doi:10.1098/rspa.2011.0384
14. Biot MA (1959) Folding of a layered viscoelastic medium derived from an exact stability theory of a continuum under initial stress. Quart Appl Math 17:185–204
15. Biot MA (1963) Surface instability of rubber in compression. Appl Sci Res 12:168–182
16. Rodriguez EK, Hoger A, McCulloch AD (1994) Stress-dependent finite growth in soft elastic tissues. J Biomech 27:455
17. Flamant A (1892) Sur la répartition des pressions dans un solide rectangulaire chargé transversalement. Compte Rendu Acad Sci Paris 114:1465–1468
18. Warne DP, Warne PG, Lee MR (2002) A summary of asymptotic finite deformation results for a point load on a hyperelastic half-space. Math Mech Solids 7:451–482
19. Mal'kov VM, Mal'kova YV (2008) Analysis of a stress singularity in a non-linear Flamant problem for certain models of a material. J Appl Math Mech 72:468–474
20. Arruda EM, Boyce MC (1993) A three-dimensional model for the large stretch behavior of rubber elastic materials. J Mech Phys Solids 41(2):389–412
21. Tanaka H, Tomita H, Takasu A, Hayashi T, Nishi T (1992) Morphological and kinetic evolution of surface patterns in gels during the swelling process: evidence of dynamic pattern ordering. Phys Rev Lett 68:2794
22. Hong W, Zhao X, Zhou J, Suo Z (2008) A theory of coupled diffusion and large deformation in polymeric gels. J Mech Phys Solids 56:1779–1793
23. Chan EP, Karp JM, Langer RS (2011) A "self-pinning" adhesive based on responsive surface wrinkles. J Polym Sci B 49:40–44

Chapter 7
Buckling Delamination of Compressed Thin Films

Myoung-Woon Moon

Abstract The aim of this chapter is to review the studies on buckle delamination in compressively stressed thin films over substrates by pulling together experimental and theoretical analysis. The general phenomena shown in delamination buckles of compressively stressed films were discussed from the onset to propagation over the substrates. The experimental observations were characterized by the delamination conditions and buckle morphologies. Then, the related mechanics for buckle delamination were provided with a theoretical solution for simple buckle configurations and a numerical solution for nonlinear buckle. Based on the experimental and theoretical analysis, the buckle configuration was applied to fluidic channels by precisely controlling buckle width within the desired area by adjusting interface adhesion.

7.1 Introduction

When the strain energy of a compressive film exceeds the interface toughness or interface adhesion energy, the film could show the interesting nonlinear behavior of buckle delamination. Nonlinear configurations of delamination buckle have been observed in many different film–substrate systems under residual compressive stress of up to several GPa, which developed from the thermal expansion coefficient mismatch or lattice parameter mismatch between film and substrate, or intrinsic stress from its bonding structure which originated during deposition [4, 8, 20]. When the film or skin is well adhered to its substrate which is relatively compliant, like polymers (e.g., PMMA, PDMS), then buckle of the thin film or skin without delamination against the substrate has been reported even under high compression

M.-W. Moon (✉)
Future Convergence Research Division, Korea Institute of Science and Technology (KIST), Seoul, Republic of Korea
e-mail: mwmoon@kist.re.kr

X. Chen (ed.), *Mechanical Self-Assembly: Science and Applications*, DOI 10.1007/978-1-4614-4562-3_7,

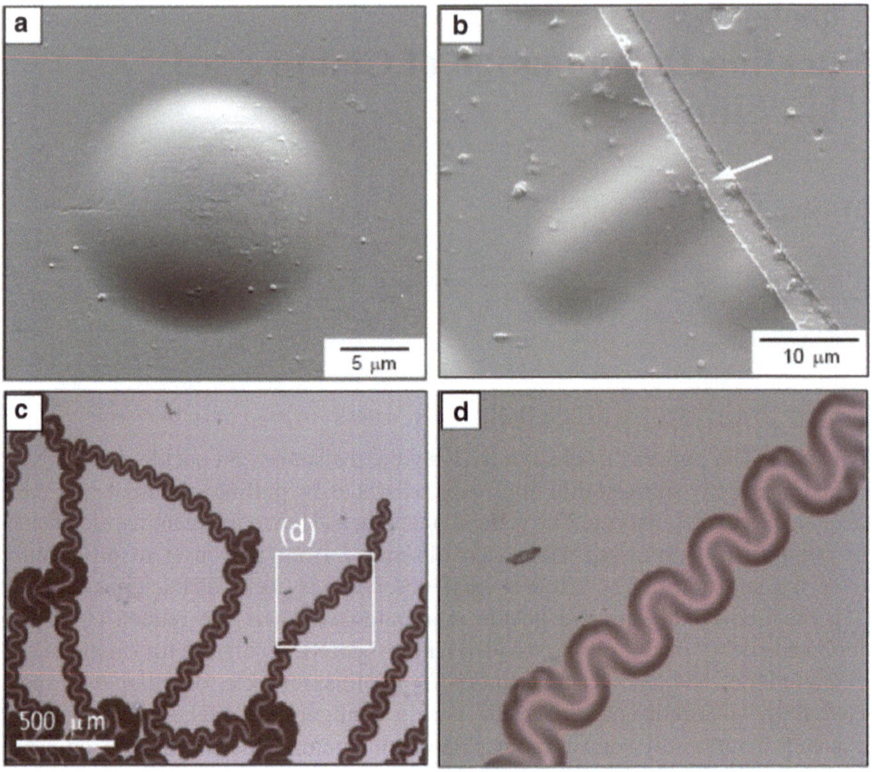

Fig. 7.1 Buckle configurations of three categories; a circular buckle (**a**), a straight-sided buckle (**b**), and a nonlinear telephone cord buckle (**c**, **d**)

[2, 3, 23]. However, on moderate substrates like glass, Si, or metal, compressively stressed films are likely to delaminate into the buckle configuration, which would further enhance the delamination [7, 19–21].

Several systems under compressive stress have been reported to reveal the buckle configuration as the stress relief pattern. With high residual stress, diamond-like carbon (DLC) films or diamond films on glass, Si, or steel have been well known for delamination buckle configurations [16, 18, 21, 29], as well as such systems as mica films glued on Al, Mo films on glass [12, 25], amorphous (hydrogenated) Si films on glass/Si [27], stainless steel on polycarbonate, Fe/Ni compositionally modulated films, Boron films on NaCl [14, 15], and Nickel films on polycarbonate substrates. Hydrogen-enhanced niobium (Nb) films on polycarbonate (PC) or mica have buckle geometries due to increased hydrogen concentration [26].

In the presence of small interface separations, the films may buckle to release the strain energy or residual stress. The ensuing buckles of engineering films on substrates exhibit several configurations, ranging from circular to linear to telephone cord as summarized in Fig. 7.1. The linear symmetric types, such as circular buckle (Fig. 7.1a) or straight-sided buckle (Fig. 7.1b) initiated along the free edge, would be rarely observed in experimental conditions due to their energetic stability, while the nonlinear and asymmetric morphology of buckle (Fig. 7.1c, d), the so-called telephone cord buckle, has been well reported in many systems [20, 22]. Note that nonlinear buckle, or telephone cord buckle, has a nonlinear side configuration, which is periodically repeated with an asymmetric unit segment. However, buckles known as straight-sided buckle and circular buckle show symmetric and linear configurations, but are rare in real experimental conditions due to their energetic instability.

In this chapter, the experimental observations were reviewed for the delamination condition and buckle morphologies. Then, the related mechanics for buckle delamination were reviewed with a theoretical solution for simple buckle configurations and a numerical solution for nonlinear buckle. The role of imperfections on the initiation and propagation of buckle-driven delaminations in compressed thin films has been demonstrated by experiments performed with DLC films deposited onto glass substrates. The surface topologies and interface separations have been characterized using the Atomic Force Microscope (AFM) and the Focused Ion Beam (FIB) imaging systems.

The profiles of several nonlinear buckles have been measured to establish the symmetry of each repeated unit, revealing similarity with a circular buckle pinned at its center. Lithographic techniques applied to a substrate prior to film deposition can create areas of low interface adhesion surrounded by regions of high adhesion. When the area of low adhesion is a strip, the width of the strip controls the buckle morphology: smooth Euler buckles for narrow strips, telephone cord buckles for wider strips, and symmetric varicose buckles under a very limited range of conditions. A complex and designed pattern of buckling delamination has been introduced for simple fluidic applications.

It has been shown that the telephone cord topology can be effectively modeled as a series of pinned circular buckles along its length, with an unpinned circular buckle at its front. Furthermore, evaluation of the energy state over the buckle unit or the energy release rate at the crack tip along the nonlinear side is conducted with 3-D numerical models.

7.2 Buckling Delamination of Thin Compressed Films

7.2.1 Imperfection-Driven Delamination Buckling

Delamination or buckle would be caused near the interface defects or free edge (Fig. 7.1) [20] with critical imperfection length due to unstable stress or strain energy. The failure responses exhibited by residually compressed thin films on thick substrates have been widely documented [5, 19, 20, 28]. Most typically, in the presence of small interface separations, the films may buckle and, moreover, the buckles can propagate beneath the film if the induced energy release rate exceeds

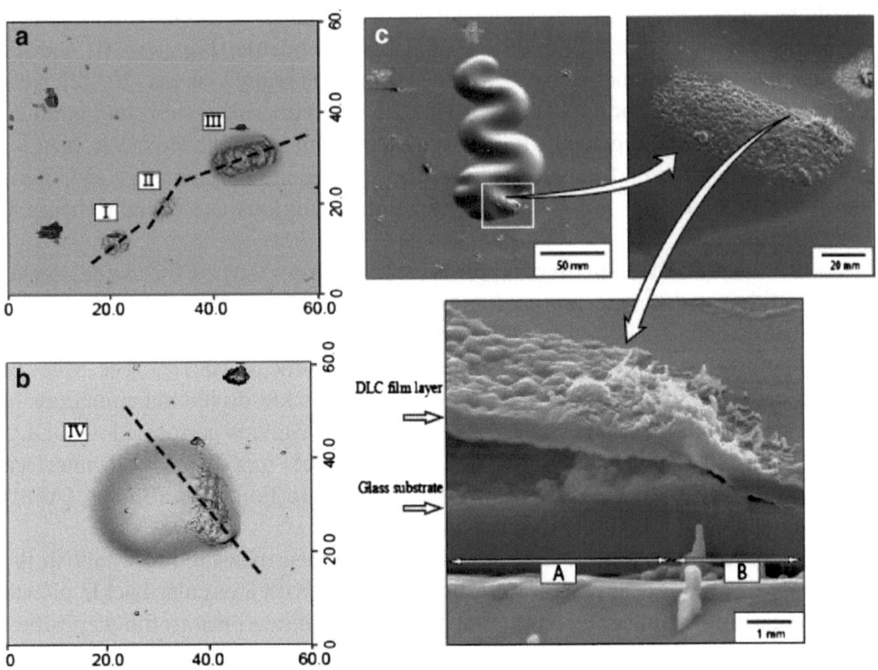

Fig. 7.2 AFM surface profiles of four imperfections. Images of three imperfections (**a**) (*I*, *II*, *III*) with small diameter and a large one (**b**) (*IV*). (**c**) SEM images with side and cross-sectional views for before and after sectioning with the FIB. Moon et al. [20], reprinted with permission

the interface fracture toughness. Since the initiation of the buckles would start from delamination of films at the interfaces that have defects, failures induced by imperfections or defects should be understood in detail.

As in all practical buckling problems, imperfections at the interfaces are expected to be important and some effects of geometric imperfections have been analyzed by either theory or experiment [11, 20]. In brief summary, the energy release rate, G, for an interface separation beneath the imperfection is dramatically altered, relative to a flat surface. Most importantly, a finite G develops even for very small initial separations and is attributed to the tensile stress normal to the interface. Measurements of the surface defects of coatings, reflected off interfacial defects, have been performed on the imperfections analyzed with AFM surface profiling and FIB cross sectioning as shown in Fig. 7.2 [20].

Briefly, the conditions for the deposition of DLC films on standard microscope slides made from soda lime glass were the following [20]. The DLC films were deposited using a capacitively coupled r.f. glow discharge, choosing conditions that generate delamination buckles. By applying a negative self-bias voltage controlled in the range of -100 to -700 V, the film thickness was in the range of 0.26–0.46 μm, and the residual compression was between about 1 and 4 GPa, resulting in delamination buckles with wavelengths of about 20–25 μm. Various surface topologies appear (Fig. 7.2a, b), indicative of a range of imperfections. It will be shown that these imperfections are all related to defects on the surface of the glass prior to DLC deposition.

AFM measurements were performed on the imperfection sites as shown in Fig. 7.2a, b. The AFM surface profiles reveal four imperfection categories. The differing profiles suggest the three basic responses shown schematically in Fig. 7.2a–c. The two smaller imperfections (numbers I and II) both have amplitudes of around 140 nm and wavelengths of about 6.2 μm. Their irregular profiles indicate that the DLC is still fully attached to the substrate, having imperfections of similar shape to those at the interfaces between film and substrate. The relatively larger imperfection (number III in Fig. 7.2a) has a substantially different profile. The irregular segment on the right (with features similar to imperfections I and II) appears to separate and buckle from the interface. The largest imperfection (number IV on Fig. 7.2b) appears at the source of a propagating telephone cord buckle. The imperfection size is determined to have amplitude, $\delta = 360$ nm and wavelength, $L_0 = 20.0$ μm.

In all cases shown above, note that the irregular topology found on the surface reflects the initial topology of the substrate, affirming that the DLC deposits everywhere with uniform thickness, $h = 0.46$ μm. Images of a larger imperfection that initiated a telephone cord buckle (Fig. 7.2c) reveal the separation. The amplitude of this imperfection exceeds the film thickness, $\delta = 1.2h$. It is also asymmetric. Region "A" on the left has a larger wavelength, $L_1 \approx 60h$. This is also the side that dictates the direction in which the buckle propagates. In region "B" on the right, the imperfection wavelength is smaller, $L_2 = 18h$, and the buckle is stationary.

7.2.2 *Buckling Delamination of Compressed Thin Films*

7.2.2.1 Linear Buckling Instability

The buckles propagate beneath the film if the induced energy release rate exceeds the interface fracture toughness. The associated mechanics have been well documented [9]. The buckles exhibit several configurations, from circular to linear to telephone cord (Figs. 7.1a and 7.3). *Straight buckles* propagate with a curved front. The conditions at the stationary side and the circular front have been modeled and rationalized in terms of mode mixity and energy release rate [9, 10, 13].

For thin films of thickness h, subject to equi-biaxial compression, σ_0, the surface displacement w normal to the substrate of a *straight-sided buckle* (Fig. 7.3c), as a function of distance y measured from the middle of the buckle well behind the curved front, is given by [9]

$$\frac{w}{h} = \frac{\xi}{2}\left(1 + \cos\left(\frac{\pi y}{b}\right)\right), \tag{7.1}$$

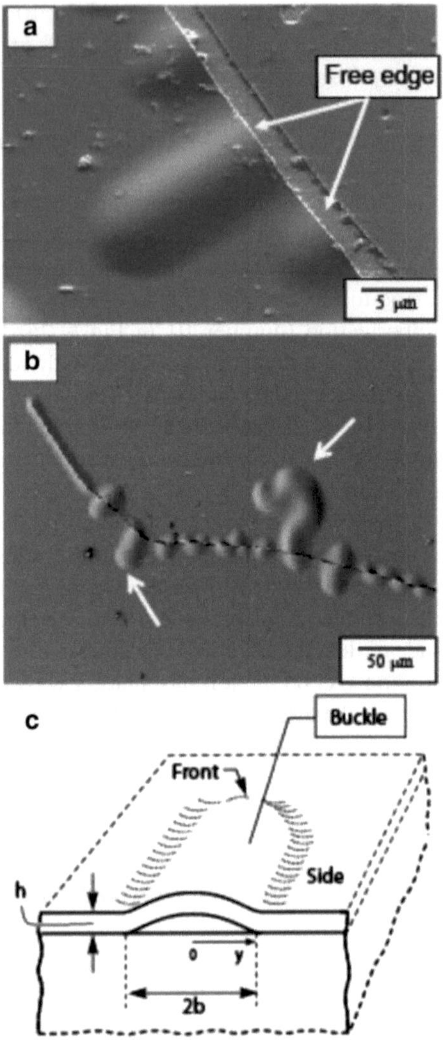

Fig. 7.3 Straight-sided buckles initiated from free edge (**a**) and buckling transition from straight to telephone cord (*arrow-marked*) (**b**). A schematic for modeling a straight-sided buckle (**c**)

where $2b$ is the width and

$$\xi \equiv \frac{w_{\max}}{h} = \sqrt{\frac{4}{3}\left(\frac{\sigma_0}{\sigma_c} - 1\right)}. \tag{7.2}$$

The critical bifurcation stress, σ_c, at width $2b$, is

$$\sigma_c = \left(\frac{\pi^2}{12}\right)\frac{E}{1-\nu^2}\left(\frac{h}{b}\right)^2 \tag{7.3}$$

with E and ν the Young's modulus and Poisson's ratio of the film, respectively. The energy release rate and phase angle, $tan\psi_s \equiv K_{II}/K_I$, *on the sides*, well behind the curved front, are

$$\frac{G_s}{G_0} = \left(1 - \frac{\sigma_c}{\sigma_0}\right)\left(1 + 3\frac{\sigma_c}{\sigma_0}\right), \quad (7.4)$$

$$tan\psi_s = \frac{4cos\omega + \sqrt{3}\xi sin\omega}{-4sin\omega + \sqrt{3}\xi cos\omega}, \quad (7.5)$$

where the energy release rate has been normalized by the strain energy per unit area when the film is released in plane strain [9]:

$$G_0 = \frac{(1 - \nu^2)h\sigma_0^2}{2E}. \quad (7.6)$$

Note that the normalized energy release rate, G/G_0, and the maximum deflection, w_{max}/h, depend only on the normalized stress, σ_0/σ_c. The phase angle, ψ, while also a function of the Dundurs' parameters [9], for present purposes is set to the value of absent elastic mismatch, whereupon $\omega = 52.1^o$.

The steady-state energy release rate averaged over *the curved front* is [9]

$$\frac{G_{ss}}{G_0} = \left(1 - \frac{\sigma_c}{\sigma_o}\right)^2. \quad (7.7)$$

The energy release rate along the sides exceeds that along the front at all σ_0/σ_c [10]. Yet, the front propagates because it experiences a significant opening mode, while the sides become exclusively mode II, once the buckle attains a characteristic width (associated with $\sigma_0/\sigma_c = 7.54$).

7.2.2.2 Nonlinear Buckling Instability

As the calculation for the nonlinear side of the stationary telephone cord buckle has been conducted, one can consider the propagation of buckles expanding through the entire film [21]. The formation mechanism of the nonlinear buckles has been considered in an experimental approach for the DLC film under the equi-biaxial stressed thin film. At the condition for the straight-sided buckle, secondary bifurcation buckles, classified as varicose-type buckles and telephone cord buckles, have been observed along the side of the preexisting straight-sided buckle. For large widths, by contrast, the telephone cord buckle, as a nonlinear instability, has only been induced with a unique zigzag configuration, which reveals the bifurcation along the curved leading front as it grows (Figs. 7.1 and 7.4).

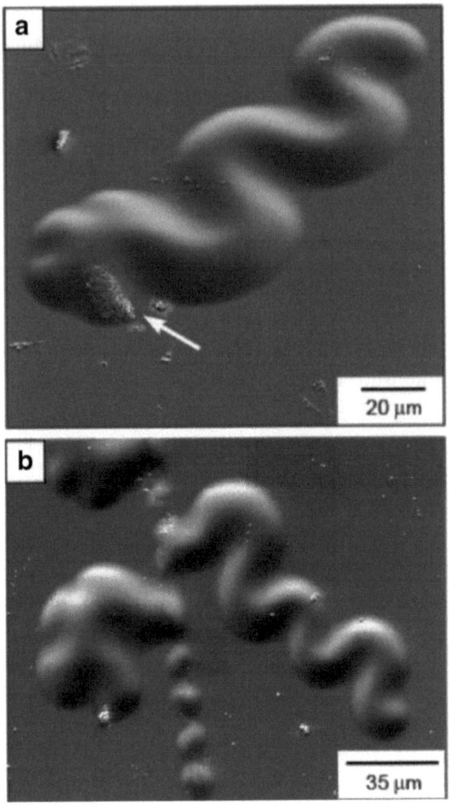

Fig. 7.4 Telephone cord buckling initiated from imperfections: interface defect (**a**) and free edge (**b**). Moon et al. [20], reprinted with permission

The two different formation sequences of telephone cord buckles have been compared with respect to buckle width. As well as the secondary bifurcation buckle from the straight side with a small perturbation on the preexisting simple straight-sided buckle with width b in the range of $2.0 < b/b_0 < 2.5$, where b_0 is the reference width for the onset of buckling, the nonsymmetric undulation has been observed with the bifurcation at the circular leading front in the natural growth range of $2.5 < b/b_0$ as the telephone cord expands. The characterization of telephone cord buckles has been provided by several equipment for measuring the exact morphologies, giving basic insight into further discussion about the energetic stability of telephone cord buckles. The axi-symmetric model for the nonlinear side of the telephone cord buckle has been compared with the straight side and circular front in terms of energy release rate and mode mixity.

The profiles of telephone cord buckles are characterized along different chords using the AFM. Images of the telephone cord buckles (Fig. 7.5) suggest that each repeated unit has a center of symmetry, denoted O in the figure, and that the circumference around that point, denoted by the arc XY, has constant curvature. The adjacent units have the inverse symmetry. Each repeated unit occupies roughly a 90° angular domain. AFM profiles measured along representative trajectories for

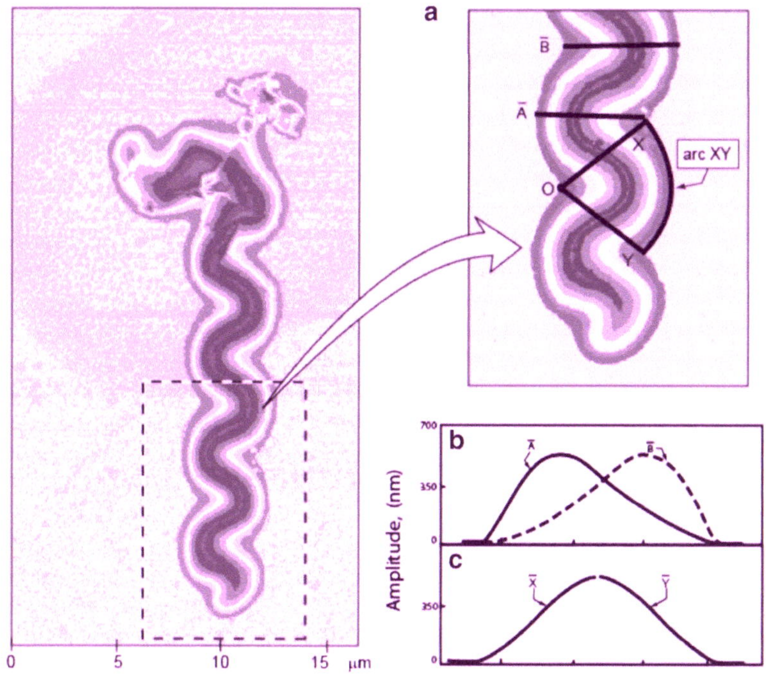

Fig. 7.5 Plan view of a telephone cord buckle, used to highlight a unit segment with the center of curvature at O and arc length XY (**a**). The profiles along the radial trajectories, A and B (**b**). The profile along the medians X and Y (**c**). Moon et al. [21], reprinted with permission

a DLC film (of thickness $h = 0.13$ μm) affirm the overall characteristics (Fig. 7.5a). *Radial trajectories* originating at O all have essentially the same asymmetric profile, exemplified by that shown in Fig. 7.5b. Note the mirror symmetry between A and B. Profiles along the medians between adjacent units (lines X and Y in Fig. 7.5a) are symmetric (Fig. 7.5c).

7.2.2.3 Energy Release Rate of Buckling

In order to estimate the energy relaxation during nonlinear buckle evolution, the full model of nonlinear buckle has been calculated for the nonlinear sides of telephone cord buckles, which reveals that telephone cord morphologies could release most of the strain energy as buckle widths or residual stress of the film increase. Besides telephone cord buckle as the nonlinear buckle morphology, the varicose type with symmetric instability along the side has also been predicted by the numerical calculation.

Given the symmetry of each unit of the telephone cord buckle, described above (see Figs. 7.1d and 7.5), it is assumed that the energy release rate and the profile can be *modeled as a full circular buckle, of radius* $r = R \equiv 2b$, *with pinned center*

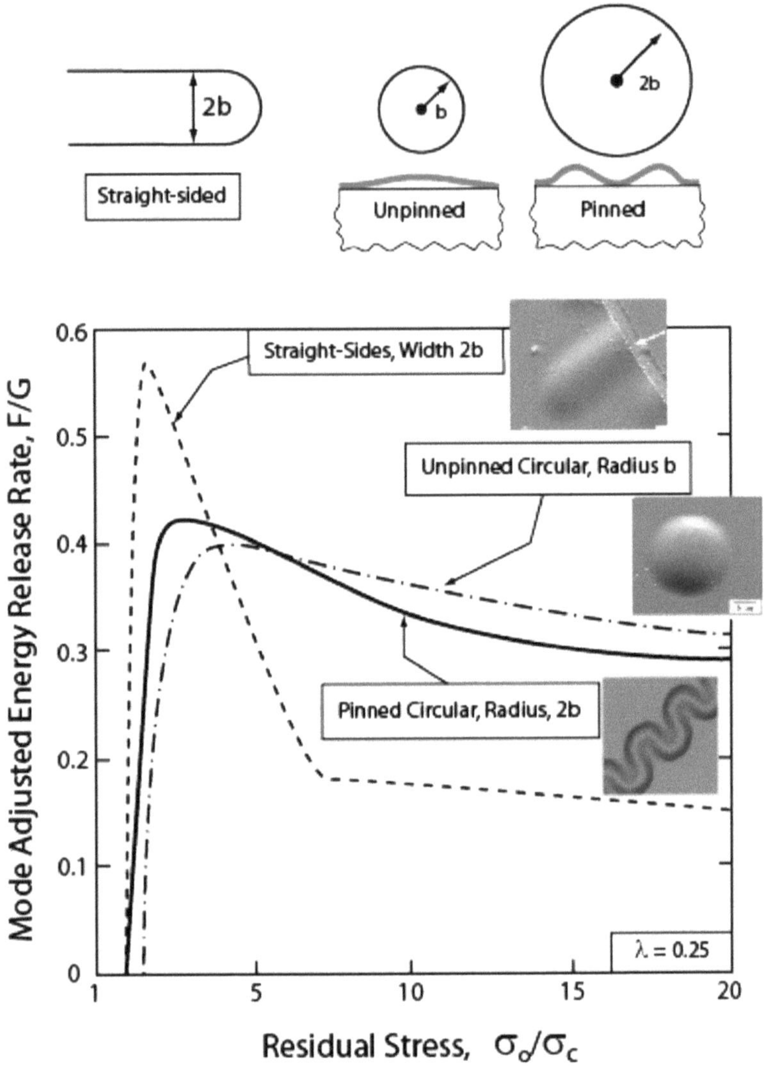

Fig. 7.6 The schematics for straight-sided (*top left* and *middle*) and telephone cord (*top right*) buckling used for analysis. Mode-adjusted energy release rate for three cases as a function of normalized film stress for three different morphologies. Moon et al. [21], reprinted with permission

(Fig. 7.6a). This assumption is validated below, using the measurements and analysis conducted for the straight-sided buckle created by the FIB. For a circular plate of radius $2b$ pinned to the substrate at the center, the corresponding critical stress is [5] $\sigma^* = 1.081\sigma_c$, and therefore only slightly greater than that of the straight-sided buckle of width $2b$.

Solutions for the circular, pinned buckle were obtained by numerical integration of the nonlinear, axi-symmetric von Karman equations [21]. The energy release rate and the phase angle are calculated using

$$G = \frac{1-\nu^2}{2Eh^3}\left(12M^2 + h^2(N+\sigma_o h)^2\right) \tag{7.8}$$

and

$$\tan\psi = \frac{K_{II}}{K_I} = \frac{\sqrt{12}M\cos\omega + h(N+\sigma_0 h)\sin\omega}{-\sqrt{12}M\sin\omega + h(N+\sigma_0 h)\cos\omega}, \tag{7.9}$$

where the bending moment is $M = [Eh^3/(12(1-\nu^2))]\mathrm{d}^2w/\mathrm{d}r^2$.

Computed results for G/G_0 and ψ for the pinned circular buckle used to model the sides of the telephone cord are presented in Fig. 7.6b as a function of σ_0/σ_c. The buckling stress, σ_c, for the straight-sided buckle of width $2b$, defined in (5.3), is used throughout to normalize the stresses. Included in Fig. 7.6b are the corresponding results for an unpinned circular buckle of radius b, obtained by numerical analysis [9, 12]. The buckling stress is $1.488\sigma_c$. To understand the trends at large σ_0/σ_c in Fig. 7.6, it is useful to identify the *total elastic energy per unit area* stored in the biaxially stressed film:

$$G_0^* = \frac{(1-\nu)h{\sigma_0}^2}{E}. \tag{7.10}$$

Note that $G_0^*/G_0 = 1.54$ for a film with $\nu = 0.3$. As σ_0/σ_c becomes large, equivalent to a large-diameter buckle, the energy release rate slowly approaches G_0^*, asymptotically releasing all the stored energy in the film. The corresponding limit for the straight-sided buckle approaches G_0, because the released film remains subject to the plane strain constraint parallel to the sides. For further assessment, it becomes convenient to express σ_0/σ_c in terms of the buckle size, b. For this purpose, a reference length is defined as the half-width of the straight-sided configuration at the onset of buckling,

$$b_0 \equiv \left(\frac{\pi}{\sqrt{12(1-\nu^2)}}\right) h\sqrt{\frac{E}{\sigma_0}}, \tag{7.11}$$

whereupon

$$\frac{b}{b_0} = \sqrt{\frac{\sigma_0}{\sigma_c}}. \tag{7.12}$$

The energy release rates for the straight-sided, pinned, and circular buckles, summarized in Fig. 7.6, indicate that G and ψ for the circular buckle and at the sides of the telephone cord are similar. By comparison, G at the sides of the straight buckle is smaller, at least when the stress is large, $\sigma_0/\sigma_c \geq 9$ (as in the present case, addressed below). Moreover, when $\sigma_0/\sigma_c \geq 7.5$, the sides of the straight buckle experience pure mode II, while the circular buckles retain a substantial component of mode I.

The existence of the telephone cord morphology is intimately related to interfaces having toughness that increases with increasing proportion of mode II to mode I. Indeed, as revealed in earlier work [9, 12], the occurrence of stable propagation owes its existence to this mode dependence. The tendency to develop a curved delamination front is tied to the larger proportion of mode I relative to mode II as the buckle enlarges.

To simulate features of telephone cords with the solutions for circular and pinned circular buckles, it is useful to introduce a phenomenological representation of a family of interface toughness dependencies [9],

$$\begin{aligned} \Gamma_c(\psi) &= \Gamma_{Ic} f(\psi), \\ f(\psi) &= \left(1 + tan^2((1-\lambda)\psi)\right), \end{aligned} \tag{7.13}$$

where $\Gamma_c(\psi)$ is the mode-dependent interface toughness, Γ_{Ic} is the mode I toughness, and λ is a mode-sensitivity parameter that sets the strength of the mode dependence. The criterion for propagation of a crack in the interface is $G = \Gamma_c(\psi)$. The ratio of mode II to mode I interface toughness values is $\Gamma_{IIc}/\Gamma_{Ic} = 1 + tan^2((1-\lambda)\pi/2)$. Interfaces with moderately strong dependence typically have $\lambda<0.3$ [6].

A mode-adjusted energy release rate, F, provides insight into the tendency of buckles to propagate on curved rather than straight edges [10]. With $\Gamma_c(\psi) = \Gamma_k f(\psi)$, let

$$F = \frac{G}{f(\psi)} \tag{7.14}$$

such that the criterion for propagation of the interface crack becomes $F = \Gamma_c$. Note that when σ_0/σ_c exceeds about 3 (or, equivalently, when $b/b_0<\sqrt{3}$), the mode-adjusted energy release rate on the straight edge is lower than that on curved sides. This behavior underlies the tendency of highly stressed films to display curved buckle morphologies. It also explains why a straight-sided buckle propagates at its curved front, rather than spreading from its straight sides.

7.2.2.4 Strain Energy for Nonlinear Buckle

Confinement of a delamination buckle to a narrow strip of low adhesion (see next section) predetermines the width of the buckle and thereby eliminates the

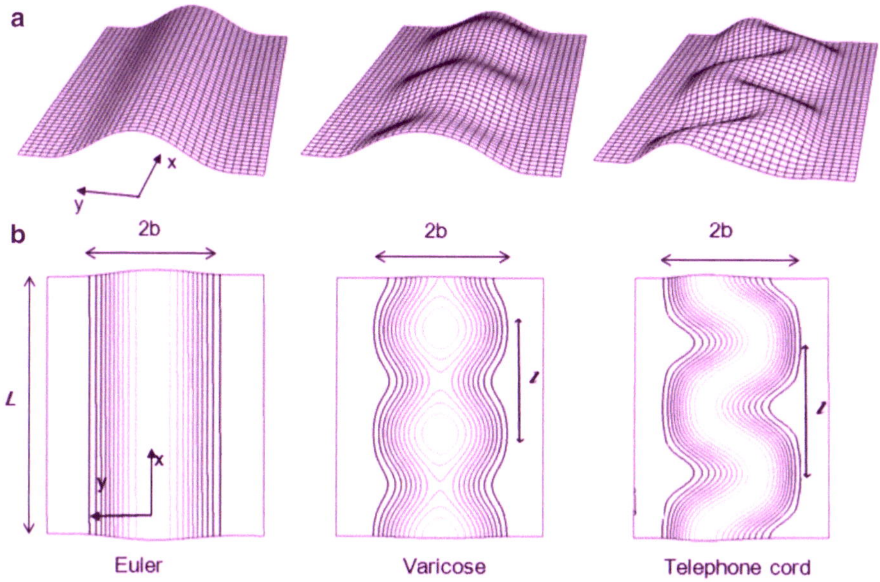

Fig. 7.7 Geometry and finite element mesh for buckling analysis of a constant width plate (film) clamped along its edges at $y = \pm b$ to a rigid substrate. The Euler mode, the varicose mode, and the telephone core mode are depicted in two views. Moon et al. [22], reprinted with permission

complicated role played by the mode dependence of the interface toughness in setting the buckle width for an unconfined delamination. The buckling and post-buckling behavior of a film under equi-biaxial compression that is detached from a substrate over a strip of width $2b$ is usually modeled as a plate of the same width that is fully clamped along it edges (Fig. 7.7). This is an excellent approximation for determining the critical stress, buckling amplitude, and relevant energy release rates as long as the Young's modulus of the substrate is not less than about one-fifth of that of the film [4, 30], as will be assumed here. When the substrate has a very low modulus compared to that of the film, deformation of the substrate along the edge of the detached region becomes important such that the assumption of a clamped edge overestimates the constraint.

The Euler mode has been used to compute the energy release rate along the sides of the buckle, but here the focus is on the energy release rate $\overline{G}$ averaged over the curved end of the interface delamination crack propagating along the strip. For a strip of low adhesion of width $2b$, steady-state conditions at the propagating end prevail once the buckle delamination is several times longer than its width. The average energy release rate is simply the difference between the energy per area in the plate, U_0, ahead of the propagating end (in the unbuckled state) and the average energy per unit area well behind the end in the buckled state, $\overline{U}$, i.e., $\overline{G} = U_0 - \overline{U}$. The energy per area in the unbuckled state of equi-biaxial compression is

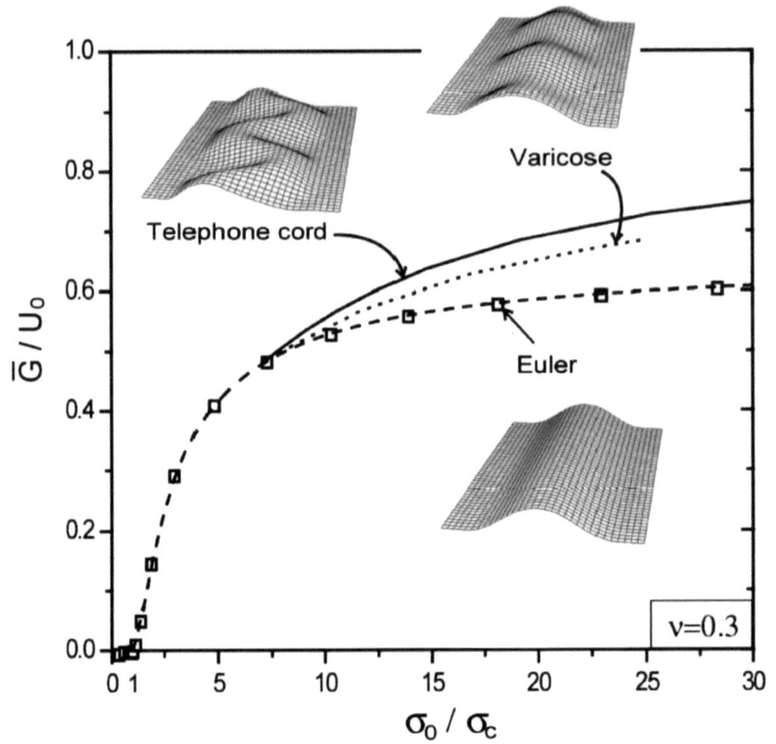

Fig. 7.8 Normalized average energy release rate for steady-state buckle delamination along a straight-sided strip, $\bar{G}/U_0$, for each of the three buckle morphologies as a function of σ_0/σ_C and b/b_0 for $\nu = 0.3$. Moon et al. [22], reprinted with permission

$$U_0 = \frac{(1-\nu)\sigma_0{}^2 h}{E} \tag{7.15}$$

while the average energy per area in the buckled state is found using (7.1)–(7.3) to be [9]

$$\overline{U} = U_0\left(1 - \left(\frac{1+\nu}{2}\right)\left(1 - \frac{\sigma_c}{\sigma_0}\right)^2\right). \tag{7.16}$$

The energy release rate for the Euler mode is [9] $\overline{G} = G_0(1-(\sigma_c/\sigma_0))^2$, where G_0 is from (7.6).

Plots of $\overline{G}/U_0$ for the Euler mode are included in Fig. 7.8, which will be introduced in the next subsection when the corresponding results for the two other morphologies are discussed. For each of the three modes, the normalized average energy release rate $\overline{G}/U_0$ is plotted in Fig. 7.8 based on the relation $\overline{G} = U_0 - \overline{U}$ or

$\overline{G}/U_0 = 1 - \overline{U}/U_0$for $v = 0.3$. By considering $\overline{G} = G_{ss}$, the relation (7.7) for the Euler mode holds for $\sigma_0/\sigma_c<6.5$ (or, equivalently, $b/b_0<2.5$) while the maximum energy release rate for $\sigma_0/\sigma_c>6.5$ is associated with the telephone cord mode. For the telephone and varicose modes, $\overline{G}$ is the energy release rate averaged over one complete wavelength of propagation. We speculate that $\overline{G}$ will asymptotically approach the elastic energy per area stored in the unbuckled film, U_0, as $\sigma_0/\sigma_c \to \infty$, but $\overline{G}/U_0$ has only attained 0.75 for $\sigma_0/\sigma_c = 30$. Note, however, that the telephone cord energy release rate $\overline{G}$ exceeds the available energy per area subject to the plane strain constraint, $G_0 = (1 + v)U_0/2$, which is the asymptotic limit for the Euler mode. Results for $\overline{G}/U_0$ for the Euler mode and the telephone cord mode are presented in Fig. 7.8b for various Poisson's ratio values, showing that this normalization $\overline{G}/U_0$ for the telephone cord mode is relatively independent of Poisson's ratio at large values of σ_0/σ_c.

7.2.3 Buckle on the Patterned Substrate

Pre-patterned regions of low adhesion at the interface between the film and substrate constrain the delamination path and thereby give rise to more predictable behavior. When the regions of low adhesion are strips, as in Figs. 7.9 and 7.10, delaminations, once nucleated, propagate the full length of the strips if the strips are above a critical width. For reasons detailed below, the Euler mode emerges for the narrow strips while a morphology much like the telephone cord mode is preferred for the wider strips. In Fig. 7.9, buckle delaminations have been nucleated at the wide delaminated region at the top of the narrow strips of low adhesion and then have propagated from the top toward the bottom of the figure. As noted in Fig. 7.9, there are several strips that are too narrow for buckling delamination to occur, i.e., those having $2b \leq 6$ μm.

Both the Euler mode and a mode similar to telephone cord mode are in evidence in Fig. 7.9, where the patterned region of low adhesion is a tapered strip. The delamination has been nucleated at the wide end of the strip and has propagated toward the narrow end where it transitions to the Euler mode. While some features of the constrained mode depart from those of an unconstrained telephone cord, in most respects the two modes are remarkably similar and the term telephone cord mode will be used here to characterize the unsymmetric undulating mode [22]. Another notable feature seen in Fig. 7.9 is the arrest of the buckle delamination at a point where its width becomes too narrow to release sufficient energy to overcome the interface toughness.

In this part, the process for buckle patterning by lithography and adhesion controlling is discussed for DLC films deposited on patterned silicon wafer substrates. Selected experimental observations of buckle delamination of patterned regions of low adhesion are presented for several types of pattern. Analysis related

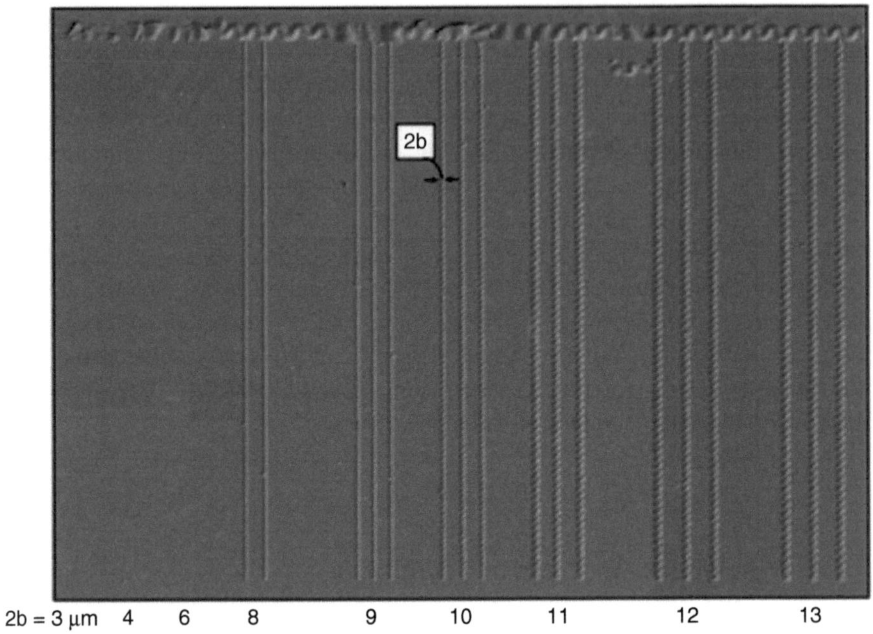

Fig. 7.9 Buckle delamination along patterned strips of low adhesion between a DLC film and a silicon substrate, showing the telephone cord morphology for wider strips and the Euler mode for narrower strips

to buckle patterning results was performed to assess strain energy and energy release rate for buckle morphologies elsewhere [24].

7.2.3.1 Patterning Experiments

The experimental procedure for patterning regions of low adhesion surrounded by regions of higher adhesion for DLC films on silicon substrates is briefly addressed with the aid of the schematic in Fig. 7.10. A 3 nm layer of Al_2O_3 between the Si and the DLC is used to create the regions of low adhesion. This layer is absent in regions of higher adhesion. The first step in creating the Al_2O_3 layer is to use E-beam lithography to expose the desired low adhesion pattern on the substrate (see Fig. 7.10). Standard lithography techniques in a clean room environment were employed. The positive E-beam resist (ER) layer was spin-coated on Si (100) followed by E-beam exposure of the regions selected for low adhesion. A very thin layer of Al is then sputter deposited which, upon oxidation, becomes an Al_2O_3 layer about 3 nm in thickness covering the low-adhesion regions. The ER layer covering the regions designated to have higher adhesion is removed with acetone and alcohol. The DLC layer is then deposited by the method of plasma-enhanced chemical vapor deposition (PECVD) using a *capacitively coupled r.f.* glow

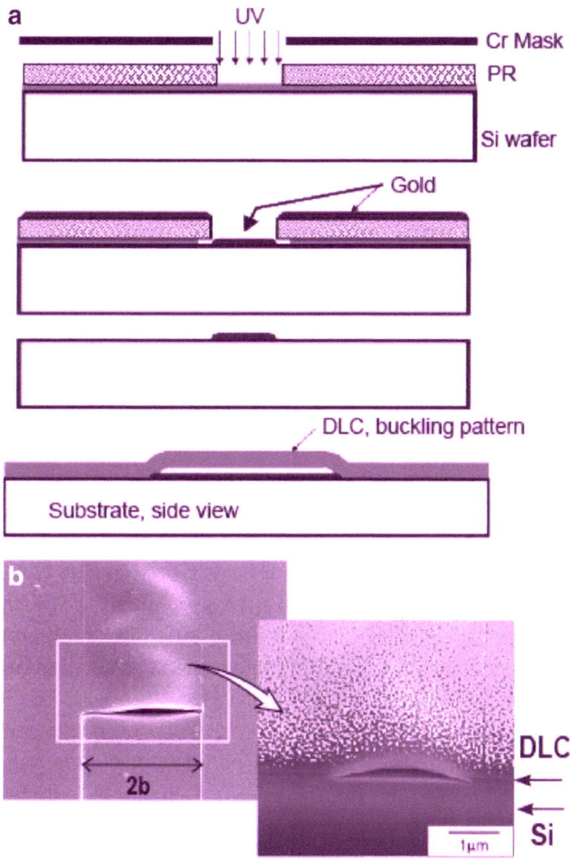

Fig. 7.10 (**a**) Schematic for creating patterns of low adhesion and (**b**) a cross-sectional view of patterned buckling delamination. Moon et al. [24], reprinted with permission

discharge, choosing conditions that from prior experience will lead to buckle delamination. Here the DLC film thickness of 0.2–0.82 μm had residual compression between about 0.9 and 2 GPa in the equi-biaxial state [4, 21]. Although no direct observation of the location of the interface crack has been made, it can be believed that separation occurs either within the Al_2O_3 layer or at the interface between the Al_2O_3 and the DLC.

7.2.3.2 Buckle Delamination on Pattern with Constant Width

The basic pattern has been classified with respect to the various shapes of release layers: linear and tapered or network. When the buckle is confined by the linear shape of the release layer (the layer that has weak adhesion), the buckle morphologies are the same along the propagation direction due to the widths being identical in Fig. 7.11.

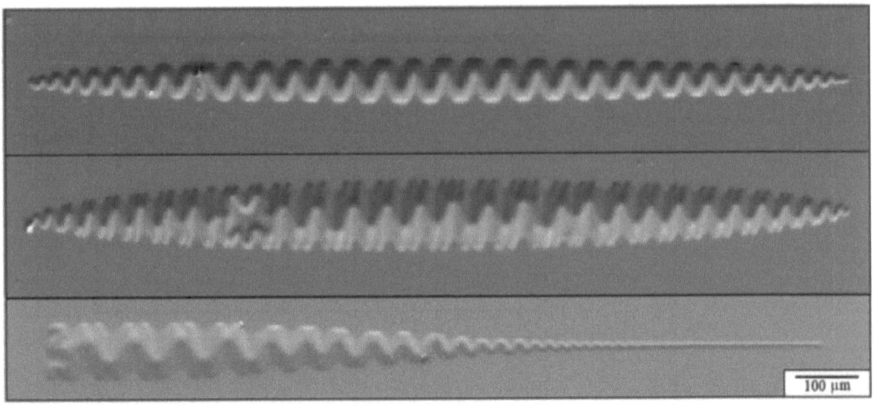

Fig. 7.11 Configuration of buckle delamination varying from straight-sided to telephone cord shape with respect to the pattern width on tapered low-adhesion layers. Moon et al. [22], reprinted with permission

Several design features of buckle patterns appear in Fig. 7.10, where the tapered release layer has been used for confinement of buckle width. As shown in the first feature in Fig. 7.10, the nonlinear side of the telephone cord buckle is unstable so that it shows the bifurcation phenomena along the nonlinear side at large width (mid of Fig. 7.10).

Regardless of buckle patterns, the configuration of buckle is shown to strongly depend on the buckle width. The array of buckle patterns shown in Fig. 7.11 could be applied for the micro (or nano) channels, which have a long tunnel inside the buckle [24]. By confinement of the buckle width to the 100 nm level, the channel with the nano-sized area could be developed, and some flows enable passing through the buckle tunnel.

7.2.3.3 Adhesion Measurement Using Buckling on Patterned Layer

Once nucleated, the condition for the buckle delamination to propagate along the strip is

$$\overline{G} \geq \Gamma_c. \tag{7.17}$$

The interface toughness, Γ_c, depends on the relative proportion of mode II-to-mode I stress intensity factors acting on the propagating interface crack front [1]. Methods to estimate the mode mix for buckle delaminations are presented in [21]. For an equi-biaxial compressive film stress σ_0, very narrow strips with $\overline{G}<\Gamma_c$ will not delaminate. However, for wider strips with $\overline{G}>\Gamma_c$, buckle delaminations will propagate unimpeded (dynamically) along the strip once nucleated. Patterning strips of various widths provide a method to bracket the interface toughness relevant to buckle delamination. *The toughness measured is that of the modified strip interface.* For the technique to be employed for measuring the toughness of a given film/substrate interface, the patterning procedure would have to be modified so as to enhance adhesion outside the strips without altering adhesion within the strip. This should be viable, at least for some interfaces, and will be pursued in subsequent work.

The DLC film on the silicon substrate in Fig. 7.10 nicely illustrates the method of bracketing Γ_c mentioned above for the strip interface. Based on the measurement techniques noted above, the DLC film in Figs. 7.9 and 7.11 is characterized with thickness, residual stress, and elastic modulus as $h = 0.26\ \mu\text{m},\ \ \sigma_0 = 1.4$ GPa, $E = 100$ GPa, $\nu \approx 0.3$. Then the onset width and the strain energies for undeformed and deformed films for (7.6) and (7.16) can be calculated to be $b_0 = 2.1\ \mu\text{m},\quad U_0 = 3.57\ \text{Jm}^{-2},\quad G_0 = 2.32\ \text{Jm}^{-2}$, respectively. The width of the widest strip in Fig. 7.9 that did not delaminate is $2b = 6\ \mu\text{m}$, which would correspond to $\overline{G} = 0.61\ \text{Jm}^{-2}$ had delamination occurred. Here the assumption is that the strip would have been wide enough to buckle had interface separation occurred, since $b > b_0$. The reason delamination does not occur is that insufficient energy is made available by interface separation, i.e., $\overline{G} < \Gamma_c$. The width of the narrowest strip that did delaminate (in the Euler mode) is $2b = 8\ \mu\text{m}$, corresponding to $\overline{G} = 1.23\ \text{Jm}^{-2}$. Thus, the interface toughness governing the buckling delamination is bracketed by

$$0.61\ \text{Jm}^{-2} < \Gamma_c < 1.23\ \text{Jm}^{-2}. \tag{7.18}$$

A more refined determination is afforded by delamination along a tapered strip.

It should be noted that the propagation condition of (7.18) is necessary but not sufficient in the sense that the fracture condition is not enforced point-wise along the curved front of the delamination, but only as an average over the propagating front. If there is a significant variation of the mode mix from point to point on the propagating crack front, identification of Γ_c with $\overline{G}$ produces the corresponding average of the interface toughness over the crack front. Point-wise estimation of the energy release rate and mode mix along the edge of a telephone cord delamination has been used to determine the interface toughness and associated mode mix of a Pt/SiO_2 interface [17]. The method was used to systematically explore the relation between one of the variables in the film deposition process and the interface toughness.

The theoretical prediction for the buckle amplitude from (7.2), i.e., $\xi = w_{\max}/h = 1.52$ or $w_{\max} = 394$ nm, compares favorably with the experimental measurement, $w_{\max} = 427$ nm.

7.2.3.4 Buckle Delamination on Tapered Pattern Layer and Grid Networks

Adhesion release layers have been designed in network patterns as shown in Fig. 7.12. On the network channel of small width in each line (Fig. 7.12a), the

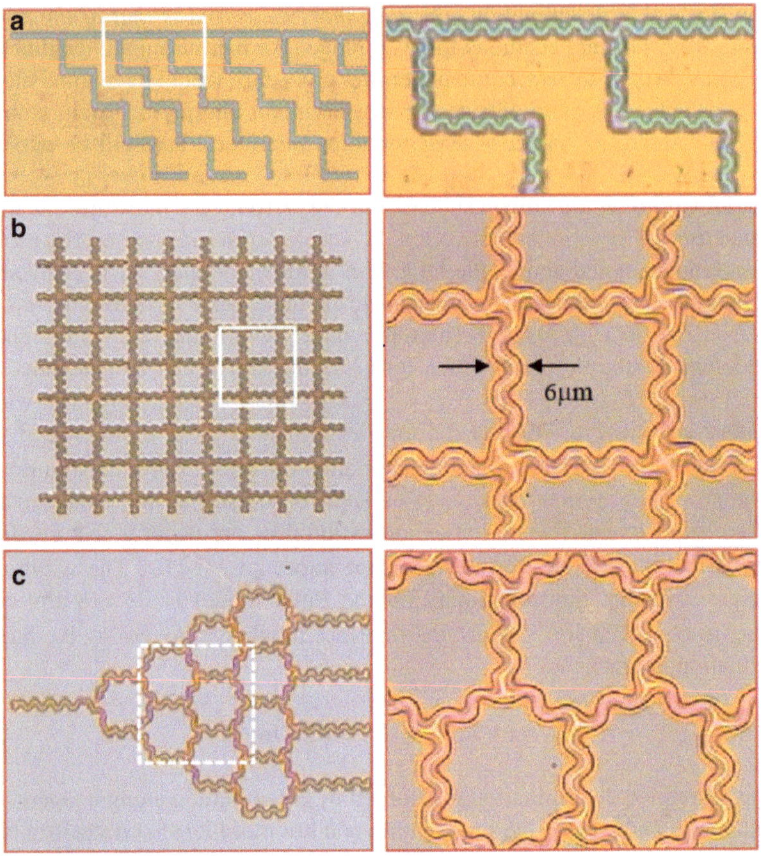

Fig. 7.12 Buckling patterns formed on various shapes of low-adhesion layer patterning. Moon et al. [24], copyright 2007 National Academy of Science, USA, reprinted with permission

straight-sided buckle grows and it passes across the next buckle branch at the junction, where the buckle is required to have fourfold symmetry, so straight-sided buckles at the junction are shown to be not stable due to there being no bifurcation mode in the Euler buckle during propagation. For the network with larger width (Fig. 7.12b), the telephone cord buckle also grows and passes across each junction. During propagation of the telephone cord buckle, it could introduce the bifurcation at the side or front so that the folding at the junction is relatively easy and it shows a stable buckle configuration at the junction as shown in Fig. 7.12b, c. In order to control and confine the buckle in a straight-sided shape at the network junctions, advanced design is required for junctions on the network.

New experiments into the control of buckle morphologies have been introduced by confinement of buckle width with interface adhesion differences. Films subject to equi-biaxial compression in the unbuckled state have been deposited on a patterned substrate. Highly controlled buckle delamination patterns and a wide variety of buckle patterns can be achieved, such as straight channel-type buckles, tapered buckles, and buckle networks. The buckle configuration has been selected with respect to the confined buckle width. The straight-sided buckle, which is rarely observed in experimental conditions, has been stably shown on patterned layers with a small buckle width, but the buckle width for the existence of the straight-sided buckle is relatively narrow. As one can see it in usual buckle conditions, the telephone cord buckle appeared on relatively large widths. On the tapered pattern, two different buckle configurations of straight sided and telephone cord are formed. The propagation behavior on complex patterns of grids and corners has been also designed.

7.3 Summary

In this chapter, delamination buckle has been fully studied in experiments and the associated theory. The initiation of buckle delamination has been discussed in terms of interface defects or imperfections, where the criterion for the critical length that induces delamination or delamination buckle was presented. Once it initiates, the buckle grows in the form of a telephone cord buckle under usual experimental conditions. The characterization of buckle configurations has been dealt with by experiments and numerical modeling. Buckle patterns were introduced by confinement of the buckle width by means of controlling the interface adhesion on patterned substrates.

The buckle delamination has been characterized by experimental analysis and modeled to obtain a theoretical estimate of the energy state. The buckle could be developed by the confinement of buckle width, which is possible by interface adhesion control. In a later work, we demonstrated the possible applications of buckles on thin films in related fields such as nano-fluidics flowing through patterned buckles [24].

References

1. Audoly B (1999) Stability of straight delamination blisters. Phys Rev Lett 83(20):4124
2. Bowden N, Brittain S, Evans AG, Hutchinson JW, Whitesides GM (1998) Spontaneous formation of ordered structures in thin films of metals supported on an elastomeric polymer. Nature 393:146
3. Chen X, Hutchinson JW (2004) A family of herringbone patterns in thin films. Script Mater 50:797
4. Cho S-J, Lee K-R, Eun KY, J-h J, Kwon D (1999) A method of determining the elastic properties of diamond-like carbon films. Diamond Relat Mater 8:1067
5. Evans AG, Hutchinson JW (1984) On the mechanics of delamination and spalling in compressed films. Int J Solid Struct 20:455
6. Evans AG, Ruhle M, Dalgleish BJ, Charalambides PG (1990) The fracture energy of bimaterial interfaces. Mater Sci Eng A126:53

7. Faulhaber S, Mercer C, Moon M-W, Hutchinson JW, Evans AG (2006) Buckling delamination in compressed multilayers on curved substrates with accompanying ridge cracks. J Mech Phys Solids (54):1004
8. Gioia G, Oritz M (1997) Delamination of compressed thin films. Adv Appl Mech 33:120
9. Hutchinson JW, Suo Z (1992) Mixed mode cracking in layered materials. Adv Appl Mech 29:63
10. Hutchinson JW (2001) Delamination of compressed films on curved substrates. J Mech Phys Solids 49(9):1847
11. Hutchinson JW, He MY, Evans AG (2000) The ratcheting of compressed thermally grown thin films on ductile substrates. J Mech Phys Solids 48:709
12. Hutchinson JW, Thouless MD, Liniger EG (1992) Growth and configurational stability of circular, buckling-driven film delaminations. Acta Metal Mater 40:295
13. Jensen HM, Sheinman I (2001) Straight-sided, buckling-driven delamination of thin films at high stress levels. Int J Fract 110:371
14. Kinbara A, Baba S, Matuda N, Takamisawa K (1981) Mechanical properties of and cracks and wrinkles in vacuum-deposited MgF_2. Thin Solid Films 84:205
15. Kinbara A, Baba S (1991) Growth process of wrinkles generated in deposited films. J Vac Sci Technol A9:2494–2496
16. Kim H-J, Moon M-W, Lee K-R, Seok H-K, Han S-H, Ryu J-W, Shin K-M, Oh KH (2008) Mechanical stability of the diamond-like carbon film on nitinol vascular stents under cyclic loading. Thin Solid Films 517:146–1150
17. Lee A, Clemens BM, Nix WD (2004) Stress induced delamination methods for the study of adhesion of Pt thin films to Si. Acta Mater 52(7):2081
18. Lee K-R, Baik Y-J, Eun K-Y (1993) Stress relief behaviour of diamond-like carbon films on glasses. Diamond Relat Mater 2:218
19. Matuda N, Baba S, Kinbara A (1981) Internal stress, Young's modulus and adhesion energy of carbon films on glass substrates. Thin Solid Films 81:301
20. Moon M-W, Chung J-W, Lee K-R, Oh KH, Wang R, Evans AG (2002) An experimental study of the influence of imperfections on the buckling of compressed thin films. Acta Mater 50:1219
21. Moon M-W, Jensen HM, Oh KH, Hutchinson JW, Evans AG (2002) The characterization of telephone cord buckling of compressed thin films on substrates. J Mech Phys Solids 50 (11):2355
22. Moon M-W, Lee K-R, Oh KH, Hutchinson JW (2004) Buckle delamination on patterned substrates. Acta Mater 52(10):3151
23. Moon M-W, Lee SH, Sun JY, Oh KH, Vaziri A, Hutchinson JW (2007) Wrinkled hard skins on polymers created by focused ion beam. Proc Natl Acad Sci U S A 104:1133
24. Moon M-W, Chung S, Lee K-R, Oh KH, Stone HA, Hutchinson JW (2007) Directed assembly of fluidic networks by buckle delamination of films on patterned substrates. Int J Mater Res 12:1203–1208
25. Ogawa K, Ohkoshi T, Tekeuchi T, Mizoguchi T, Masumoto T (1986) Nucleation and growth of stress relief patterns in sputtered molybdenum films. J Appl Phys 25:695
26. Pundt A, Nikitin E, Pekarski P, Kirchheim R (2004) Adhesion energy between metal films and polymers obtained by studying buckling induced by hydrogen. Acta Mater 52:1579
27. Thouless MD (1993) Combined buckling and cracking of films. J Am Ceram Soc 76(11):2936
28. Tolpygo VK, Clarke DR (2000) Spalling failure of α-alumina films grown by oxidation: I. Dependence on cooling rate and metal thickness. Mater Sci Eng A278:151
29. Weissmantel GHR, Schurer C, Frohlich F, Grau P, Lehmann H (1979) Mechanical properties of hard carbon films. Thin Solid Films 61(2):L5
30. Yu HH, Hutchinson JW (2002) Influence of substrate compliance on buckling delamination of thin films. Int J Fract 113:39

Chapter 8
Delaminated Film Buckling Microchannels

Alex A. Volinsky and Patrick Waters

Abstract This chapter describes the method of manufacturing microfluidic microchannels formed by delaminated buckled thin films. Thin films under compression tend to delaminate and buckle. Microchannel geometry can be controlled by tailoring film residual stress and placing patterned adhesion-weakening layers utilizing photolithographic techniques. Results based on the photoresist as the adhesion weakening layer and compressed tungsten thin films are described along with the corresponding thin film mechanics.

8.1 Introduction

Buckling delamination blisters are commonly observed in compressed thin films. These mechanically active features arise from a local loss of adhesion between the film and the substrate, due to the relief of residual stress, and exhibit directional growth. Thin films in compression can also develop circular, asterisk, straight, and phone-cord blisters. "Telephone cords" are wormlike buckled features observed in thin films, and are called so because they look like a twisted phone cord when they are viewed under an optical microscope (Fig. 8.1). The geometry and path of these blisters can be controlled by placing adhesion-weakening layers, thereby outlining the desired in-plane blister configuration and forming a channel network on a chip.

Microfluidics, as a field, has been growing with the new advances in nanotechnology [1]. This field is estimated to grow at an exponential rate, and is expected to continue to do so in the decades to come. This relatively new technology has different possible applications, from sensors to drug delivery. It relies on the ability to transfer fluids at small-scale chip sizes. Recently these were successfully used to

A.A. Volinsky (✉) • P. Waters
Department of Mechanical Engineering, University of South Florida,
4202 E Fowler Ave ENB118, Tampa, FL 33620, USA
e-mail: volinsky@usf.edu

X. Chen (ed.), *Mechanical Self-Assembly: Science and Applications*,
DOI 10.1007/978-1-4614-4562-3_8, © Springer Science+Business Media New York 2013

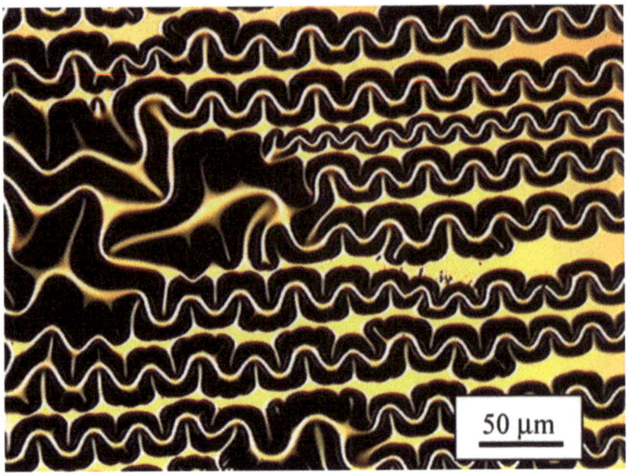

Fig. 8.1 Telephone cord buckling delamination of a compressed W film

deliver pharmacological agents directly to wound sites [2]. The channels used are commonly etched in silicon or other materials by means of standard methods of lithography. While it is not problematic to form an open trench, forming a channel is more complicated. There are multiple steps involved in the conventional technology [3], so the whole process is cumbersome, expensive, and labor intensive. In addition, Si is not biocompatible. Although one could envision a Si-based lab-on-a-chip device, the Si substrate has to be well isolated from the biological species and environment.

An alternative technology employs thin film delamination blisters which can form open micro- and nano-channel networks of specific geometry. Telephone cord delamination morphology is commonly observed as a result of the relief of residual compressive stresses within the thin film via interfacial debonding. Here, the biaxial film stress is partially relieved by film buckling in the direction perpendicular to the telephone cord propagation and by "secondary" blister buckling in the direction parallel to the telephone cord propagation. The combination of these types of buckling results in the observed sinusoidal fracture patterns [4, 5]. Normally, telephone cord blisters "run out of steam" and stop once the interfacial toughness exceeds the strain energy release rate. It is possible to make blisters propagate further by either putting mechanical energy into the system [5] or introducing liquids to the crack tip [6]. Inspired by recent results of moisture-induced thin film fracture [7], the thin film buckling delamination phenomenon was utilized as a way of manufacturing channel networks for the purposes of fluid transport in lab-on-a-chip devices and other microfluidic applications. In addition to forming channels, the thin film buckling delamination can act as a pump, thereby transporting the fluid during delamination. This allows for the enlargement of the channel cross section by using etching solutions during manufacturing.

Thin film and coating buckling delamination are truly multi-scale phenomena, with features possibly ranging from tens of nanometers to centimeters in width. In addition, compatibility with conventional microelectronic processing is very promising from the standpoint of full-scale on-chip integration. Crucial elements of a lab-on-a-chip device are channels capable of transporting the fluids within the chip dimensions and an interface which is manipulatable from the outside world. Channels allow various fluids to be mixed, stored, and passed through. Typical fluids include blood, protein solutions in various buffers, cell suspensions, etc. Several measurements can be performed by microfluidic devices, including fluid viscosity and pH [8, 9].

The competitive advantage of the channel miniaturization is due to the fact that the Reynolds number is low, allowing for laminar flow to prevail and convective mixing of the fluids not to occur. Miniaturization of the channels, and their integration onto a single chip, would allow for the handling of minute amounts of sample material, thus increasing the sensitivity. The analyte concentration A_i is inversely proportional to its volume, V, and is given in terms of the device efficiency, η_s, as in [1]

$$A_i = \frac{1}{\eta_s N_\mathrm{A} V}, \tag{8.1}$$

where N_A is the Avogadro's number. Reduction of the sample size provides higher device sensitivity, as well as the capability of handling smaller samples. One can easily and effectively perform immuno- and DNA probe assays by using minute amounts of sample material.

In addition, the channel profile geometry has a substantial influence on the fluid flow. It is important to know the distribution of the analytes in the channel for developing functional and reproducible assays. One would want to increase the molecule residence time for quantitative time-dependent analyses. The prediction of the molecule residence time is a complicated task, as it depends on the molecule location in the channel profile. A low-aspect-ratio channel velocity profile allows lateral diffusion to occur in the vertical dimension without an accompanying change in the residence time of the diffusing molecules. Therefore, one would desire microchannels with low aspect ratios for accurate quantitative studies. Thin film buckling channels have a low aspect ratio, i.e., height-to-width ratio.

8.2 Thin Film Buckling Delamination

Many industries, including microelectronics and microfluidics, depend on well-adhered thin films. Thin films are normally stressed, with biaxial residual stress residing in unpassivated films [10]. Stress in a thin film typically causes substrate bending. This effect is employed in measuring the macroscopic residual stress found within thin films by using Stoney's equation [11]. There are many different stress relief mechanisms observed in thin films. The residual stress can be partially relieved by plastic deformation and surface reconstruction. It may also be more completely relieved by mechanical film failure, especially if the stress levels are high and/or externally applied stresses are present.

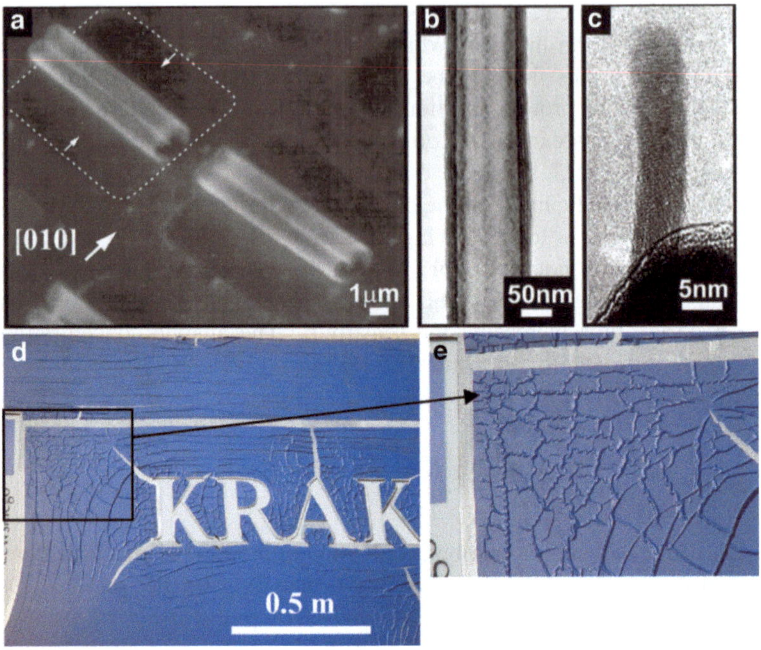

Fig. 8.2 (**a–c**) SEM and HRTEM images of InGaAs/GaAs nanotubes rolled up from bi-layered films: (**a**) Two scrolled single walls; (**b**) multi-wall nanotube (6 turn, film thickness: 4ML GaAs +4ML InGaAs); (**c**) single-wall nanotube formed from 2ML GaAs + 1ML InAs. Reprinted with permission from [23, 24]; (**d**) and (**e**) Polymer film buckling delamination on a street sign due to seasonal temperature variations. Waters and Volinsky [31], reprinted with permission

Telephone cord delamination has been observed in various thin film systems, including tungsten, carbon, diamond-like carbon (DLC), TaN, and other films which have been applied upon various substrates [12–18]. Some theoretical considerations of the formation of the telephone cord buckling geometry can be found in [19–22]. Most of the time delamination blisters run parallel to each other in blanket, non-patterned films (Fig. 8.1). The multi-scale nature of these phenomena can be seen in Fig. 8.2a–e. Figure 8.2a–c shows ultrathin films (several monolayers—ML) rolling into nanotubes [23, 24], while Fig. 8.2d–e shows polymer film buckling delamination on a street sign, several meters long, induced by seasonal temperature fluctuations, thereby causing the film buckling and adhesion loss. Telephone cord delamination morphology is commonly observed as a result of the thin film residual compressive stress relief by interfacial debonding (Fig. 8.3).

If the residual stress is less than four times the buckling stress, straight blisters become possible under biaxial residual compressive stress. However, this level of residual stress may not be high enough to initiate delamination. That is why residual stress-induced straight blisters are rarely observed in thin films. This situation changes, however, when the stress is applied externally, or when the adhesion is somehow weakened. A simple experiment with a compact disc (CD) in 3-point

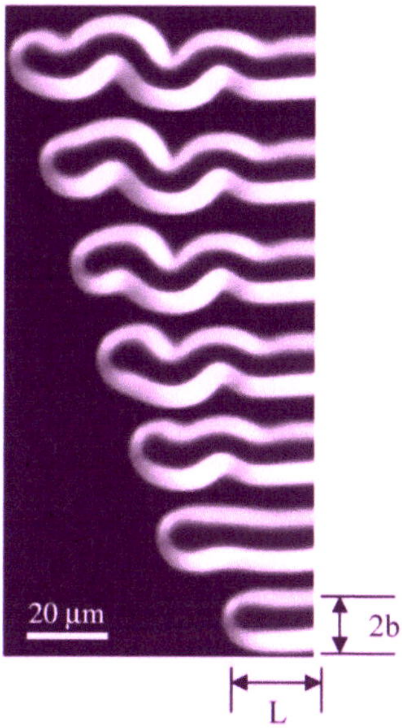

Fig. 8.3 Transition from straight-sided blister to sinusoidal shape

bending has been described in [4] to support this. Detailed dynamics of real-time telephone cord blister propagation can be observed online [25]. As was previously stated, telephone cord blisters "run out of steam" and stop once the interfacial toughness exceeds the strain energy release rate. It is possible to make blisters propagate further by either putting mechanical energy into the system [5] or introducing liquids at the crack tip [6, 7]. Liquids can also provide the mechanical energy (through surface tension and capillary forces) needed to continue the blister propagation and can also reduce the film interfacial toughness. These effects were attributed to environmentally assisted cracking, which caused the thin film delamination, and is similar to a commonly known example of the degradation and fading of vintage mirrors, as well as solar mirrors [26]. If one were to use a substrate etching solution as a fluid, then, in addition to causing thin film delamination, it would etch the substrate, thereby enlarging the channel cross section.

Tests have been conducted that demonstrate the method of fluid introduction as a means of propagating delamination blisters. These blisters continue to propagate until the fluid has been removed, or until the delamination reaches the end of the sample [6]. All of the samples contained a 1 μm thick W superlayer deposited on either thin magnetic layers or 40 nm thick Cu films on Si substrates, with a thermally grown SiO_2 layer. Figure 8.4 shows two snapshots, taken 3 min apart, illustrating water-induced blister propagation. The evidence of fluid transport is presented in Fig. 8.5. Here, the buckling delamination propagated from the right to the left, and the water, upon reaching the far left side of the sample, moved down the edge of the Si wafer. This demonstrates one of the potential mechanisms of fluid transport via blister propagation. In this case, water was transported, with the advance of the crack tip, at about 10 μm/min, although higher propagation rates,

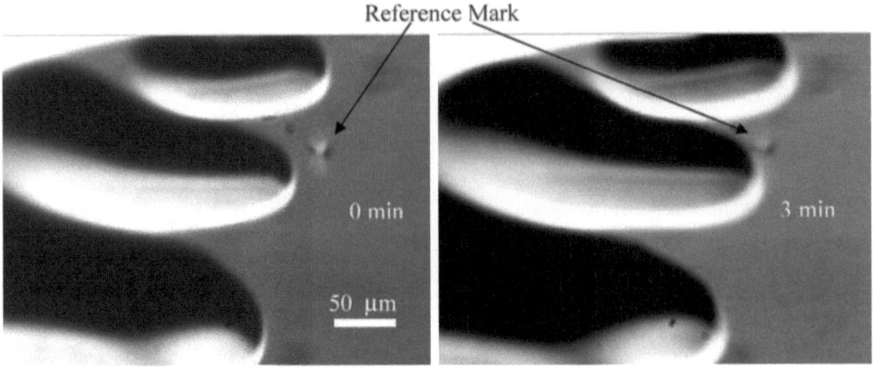

Fig. 8.4 Water-induced blister growth. Volinsky et al. [7], reprinted with permission

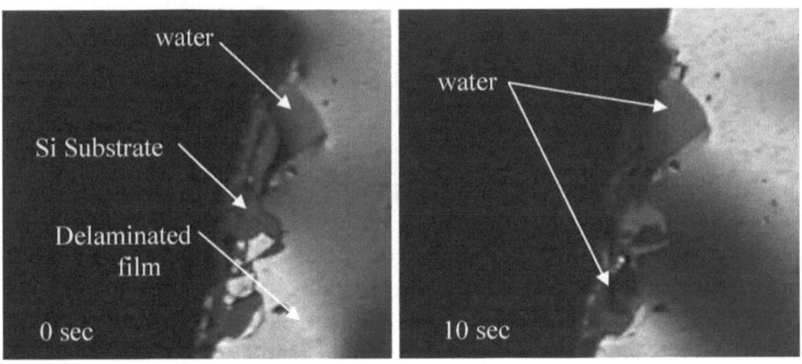

Fig. 8.5 Water transport in the delaminated blister. Volinsky et al. [7], reprinted with permission

up to several microns per second, were observed in other samples. It has been previously shown that interfacial cracks in thin films act as a vacuum cleaner, collecting contaminants from the atmosphere, thus reducing the energy of the newly formed fracture surfaces [13, 14]. In this case, the crack is acting like a pump, thus transporting the fluid along with the blister propagation. Water droplets can be seen on the Si substrate in Fig. 8.5.

In order to make a microfluidic or lab-on-a-chip device, the placement of delamination fluid transport channels needs to be exact and repeatable. Standard lithographic techniques have demonstrated the ability to control the blister placement and the buckling geometry [27]. Both telephone cord as well as straight blister geometries are available and depend on the film stress level and the width of the adhesion-weakening (release) layer [27]. With the use of moisture, the requirements on the release layer adhesion properties are less stringent, as the stress combined with the moisture is sufficient to cause delamination. The properties of the internal walls of the channel can be altered by varying the underlying materials. For example, it has been shown that fracture in low-K dielectric films is

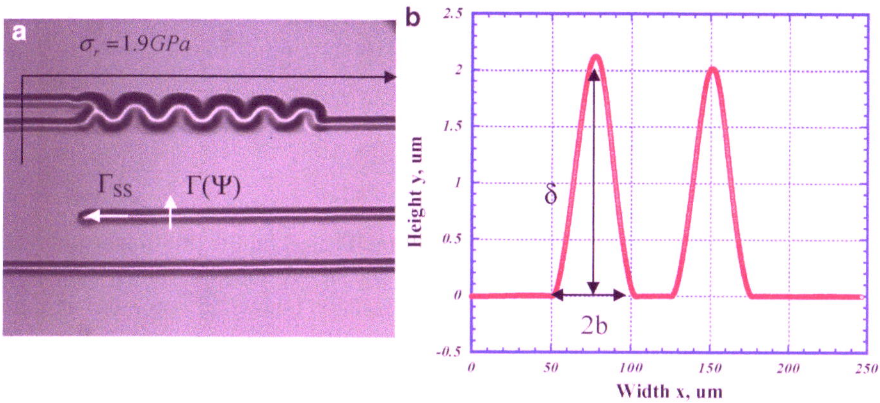

Fig. 8.6 (**a**) Optical micrograph of two straight buckled channels coalescing into one through a telephone cord delamination, and (**b**) corresponding height profiles. Volinsky et al. [7], reprinted with permission

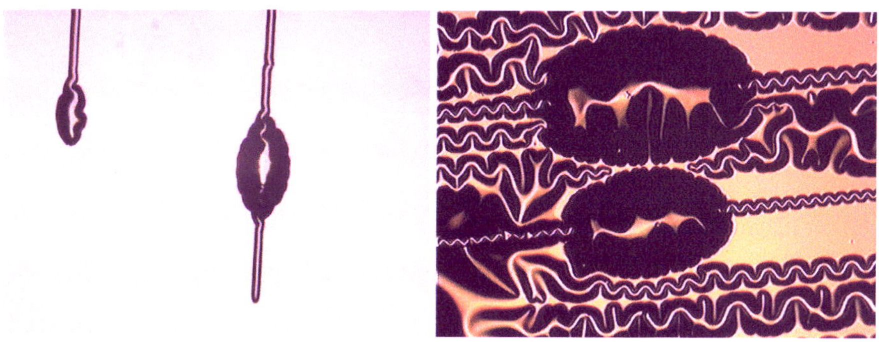

Fig. 8.7 Buckling delamination showing channels and fluid "storage reservoirs"

cohesive, i.e., the crack propagates in the film itself [28]. A similar effect is expected for a polymer release layer and it may even be possible to use regular photoresistant "lining" inside the channel. One could also use a biocompatible material as an adhesion-weakening layer, which would include polymer and ceramic films, as well as DLC.

Figure 8.6a shows two 50 μm wide straight channels, which coalesce by forming a telephone cord delamination blister, resulting in a single 50 μm wide channel. The profilometer scan in Fig. 8.6b reflects the blister buckling geometry. A highly desirable microchannel aspect ratio of 1:25 is achieved here. This demonstrates the possibility of constructing a fluid-mixing device by employing thin film buckling channels. In this particular case, the double-to-single channel transition occurred naturally, although the same result can be achieved with patterned release layers, which form channel network structures. Another important example of a storage element is shown in Fig. 8.7, where a channel extends to a larger reservoir.

One can envision using this storage element as a pump, when combined with an electrostatic device capable of mechanically pressurizing it from the top.

8.3 Buckling Blister Mechanics

The buckling channel geometry can be predicted based on the multilayer material properties, residual stress, and adhesion [22]. The steady-state interfacial toughness of the film, in the direction of the blister propagation (shown in Fig. 8.6a), is

$$\Gamma_{SS} = \frac{(1-\nu^2)h\sigma_r^2}{2E}\left(1-\frac{\sigma_B}{\sigma_r}\right)^2, \tag{8.2}$$

where σ_r and σ_B are the residual (compressive) stress and the buckling stress, respectively, h is the film thickness, E is the elastic modulus, and ν is the Poisson's ratio. The buckling stress is given in terms of the film thickness and the blister width, b, as

$$\sigma_B = \frac{\pi^2}{12}\frac{E}{(1-\nu^2)}\left(\frac{h}{b}\right)^2. \tag{8.3}$$

Now one can express the residual stress in terms of the buckling stress and the blister heights, δ, as

$$\sigma_r = \frac{3}{4}\sigma_B\left(\frac{\delta^2}{h^2}+1\right). \tag{8.4}$$

To complete the formulation, the mode-dependent interfacial film toughness, in the buckling direction, perpendicular to the blister propagation, can be expressed as

$$\Gamma(\Psi) = \frac{(1-\nu^2)h}{2E}(\sigma_r-\sigma_B)(\sigma_r+3\sigma_B). \tag{8.5}$$

This simple analysis basically shows that the level of the residual stress and the extent of the thin film adhesion will control the microchannel geometry. The upper layer residual stress, determined primarily by the deposition parameters, will control the microchannel height.

Typically, residual stress in thin films results in negative consequences that include wafer bowing, film cracking, and delamination. However, depositing a film with a compressive residual stress has been demonstrated to be a benefit, specifically when using superlayer indentation to determine the thin film interfacial toughness [12]. A similar concept can be employed to create microchannels. When the strain energy release rate, G, exceeds the interfacial toughness, Γ_i, of a film/

substrate, delamination will occur. A simplified form of the strain energy release rate in a stressed film is [22]

$$G = Z\frac{\sigma_f^2 h}{E_f}, \tag{8.6}$$

where σ_f is the stress in the film, h is the film thickness, E_f is the modulus of elasticity, and Z is a dimensionless parameter that depends on the geometry.

In thin film systems with biaxial compressive stresses, various shapes of the delaminated regions will arise. These shapes include long straight-sided, circular, and telephone cord delamination, which is the most commonly observed morphology. The delamination shape and size will depend on factors such as the film stress, thickness, and interfacial toughness. Most importantly, the propagation depends on the interfacial toughness, which increases as the mode mixity, acting on the interface ahead of the delamination crack, shifts from mode I towards mode II. Interfacial failure starts with the film delaminating from the substrate and is followed by the spreading of the delamination as the buckling loads the edge of the interfacial crack, resulting in a failure phenomenon that couples both buckling and interfacial crack propagation.

To create useful microchannels via film delaminations, the direction and morphology of the delaminations need to be controlled. The easiest way to control delamination is by controlling the interfacial toughness. This can be done by creating adhesion-reducing layers that have a lower interfacial toughness than the surrounding areas. A compressively stressed film can be used in conjunction with a patterned adhesion-reducing layer as a method for creating microchannels. Using delaminations will hopefully solve the existing challenges and provide an alternative method in creating microchannels.

8.4 Buckling Microchannel Pattern Manufacturing

To create the adhesion-reducing layers, standard photolithography techniques were employed [29]. The simplest approach used the patterned photoresist as the adhesion reducer. The basic steps in lithography include the following:

(1) The application of a photosensitive material (photoresist)
(2) Soft bake of the photoresist
(3) Exposure of the photoresist
(4) Development of the exposed pattern
(5) Hard bake of the remaining pattern

Figure 8.8 is a depiction of the basic lithography steps used here, along with the additional step of depositing a compressively stressed film for creating delamination-induced microchannels.

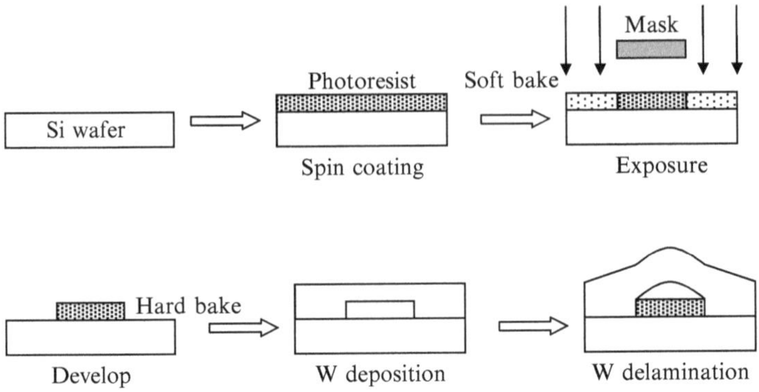

Fig. 8.8 Process of creating microchannels

There are two types of photoresist to choose from: positive and negative photoresist. Positive photoresist is exposed to UV light wherever it is to be removed. Regarding the positive photoresist, exposure to the UV light changes its chemical structure so that it becomes more soluble in a developer, which is opposite to that of the negative photoresist. The exposed photoresist is then washed away by a developer solution, leaving areas of the underlying material bare. Therefore, the mask contains an exact copy of the pattern, which is to remain on the wafer. In general, positive photoresists provide clearer edge definition than negative photoresists. The better edge definition found in positive photoresists makes them a better option for high-resolution patterns.

Negative photoresists behave in the opposite manner. Exposure to the UV light causes the negative photoresist to become polymerized, and consequently more difficult to dissolve. Therefore, the negative photoresist remains on the surface wherever it is exposed, and the developer solution removes only the unexposed portions. Thus, masks used for negative photoresists contain the inverse of the pattern to be transferred. Shipley 1813 was used here, which is a positive photoresist that is optimized for G-line exposure (436 nm wavelength).

Normally, to start the photoresist application process, the wafer surface is prepared in a specific way in order to remove the surface moisture and other contaminants. In order to remove the moisture, the wafers are baked and then primed with an adhesion promoter. Hexamethyldisilazane (HMDS) is normally used as the adhesion promoter and is applied at a reduced pressure to form a monomolecular layer on the wafer surface, thereby making the wafer hydrophobic, which prevents moisture condensation. These wafer preparation steps were ignored here because the goal of the photoresist layer was to be an adhesion reducer and only a proof of concept was intended.

Spin coating was used to apply the photoresist with thicknesses ranging from a few hundred nanometers to a few microns. If thicker coatings are required, electrochemical coatings, spray coatings, and casting processes can be used [29]. A Laurell Technologies WS-400A-8NPP/Lite Spin Processor was utilized here and can handle wafers up to 8″ in diameter. It uses a vacuum chuck to hold the

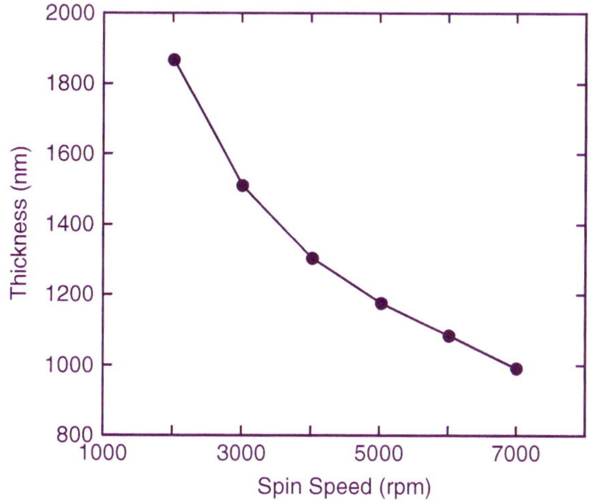

Fig. 8.9 Photoresist thickness as a function of spin speed for Shipley 1813

wafers in place and has a spin speed which can range from 0 to 6,000 rpm. Acceleration profiles, speed changes, and spin times can all be programmed by the user. By adjusting the spin speed, the photoresist thickness can be controlled. The manufacturer-supplied thickness versus spin speed data for the Shipley 1813 is shown in Fig. 8.9. They suggest that the Shipley 1813 has the best coating uniformity between spin speeds of 3,500 and 5,500 rpm.

The spin speed and time used here were 4,000 rpm for 40 s, with a ramp speed of 1,000 rpm/s. Based on the manufacturer's supplied data for the photoresist thickness versus spin speed, it was expected to have a photoresist thickness of approximately 1.3 μm after application. However, the resist thickness was measured to be approximately 1.5 μm with relatively good thickness uniformity across the wafer diameter. Spin-on deposition of the photoresist was a tricky process because the photoresist was applied manually. First, the wafer surface had to be thoroughly cleaned of any debris to avoid streaking of the photoresist during spinning. After the surface cleaning, the wafer had to be centered on the vacuum chuck to avoid uneven distribution of the photoresist. A drop of photoresist was then placed in the center of the wafer and a few additional drops were added during the first 5 s, in which the spin speed of the wafer was ramped to the maximum rpm. Upon completion of the spin-on deposition, the wafer was placed on a hot plate for 90 s at 90 °C.

After the photoresist soft bake, a range of exposure times were attempted in order to find the optimum photoresist thickness. The light source used for the photoresist exposure was a mercury vapor lamp, which provides a wavelength spectrum from 310 to 440 nm. A Karl Suss MA 56 Mask Aligner was used in hard contact mode for the mask and wafer alignment. The mask had a repeating pattern, which consisted of two straight lines running parallel, where the line width was measured to be 250 μm with 30 μm separation. The results of the photoresist profiles, after testing a range of exposure times, can be seen in Figs. 8.10 and 8.11. The photoresist width and thickness varied by up to 10% across the 4″ wafer diameter.

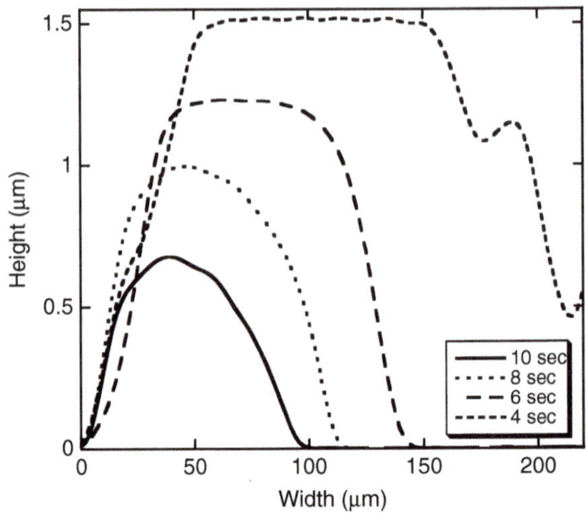

Fig. 8.10 Photoresist profiles after different exposure times

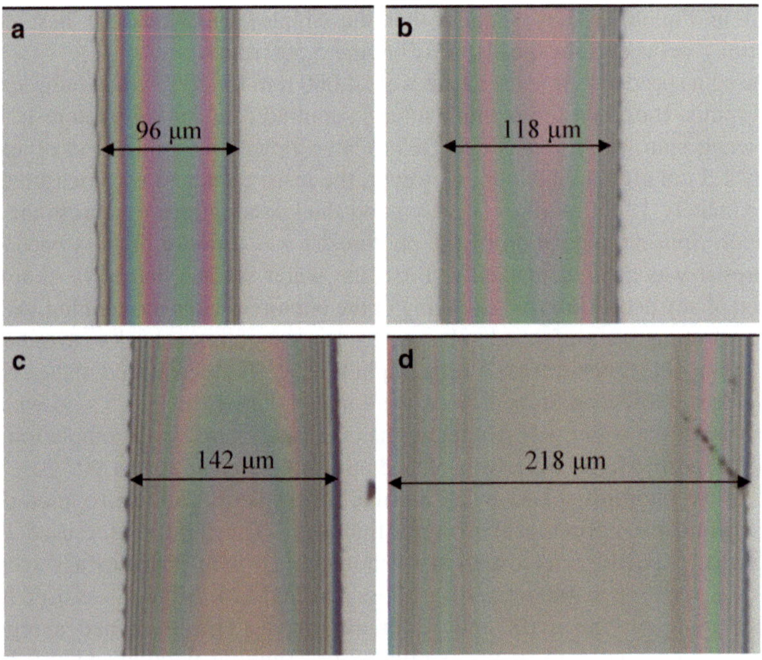

Fig. 8.11 Effect of exposure time on the final photoresist thickness: (**a**) 10 s, (**b**) 8 s, (**c**) 6 s, and (**d**) 4 s

At the shortest exposure time of 4 s, the remaining developed photoresist measured 1.5 μm in height and 218 μm in width. However, on most of the wafer there also remained a thinner layer of photoresist between the parallel lines. It was concluded that an exposure time of 4 s at 275 W was inadequate. From there the exposure time was increased by 2-s increments, with the final resist height and width being inversely proportional to the exposure time. Along with decreasing line height and width, the resist profile ended up being more rounded as the exposure time was increased.

After exposure, the wafers were placed in a developer bath for 30 s and were then rinsed with deionized water. Nitrogen was used to remove the majority of the excess moisture on the wafer, which was then hard baked on a hot plate at 110 °C for 100 s.

Overall, the lithography techniques used here were adequate in accomplishing the objective of creating simple patterns on a wafer. However, a more thorough approach can be taken in the future for optimizing spin speed and time, bake time and temperature, and exposure and development time. It must be understood that the line widths created here are relatively large when compared with other lithography techniques. Therefore, the quick optimization steps used here were adequate. If smaller line widths and heights were required, a more meticulous approach would be needed. The final step in the process was to deposit a tungsten film that is forced to have compressive residual stress.

8.5 Buckling Delamination Morphology

The various shapes of buckled regions evolve in film/substrate systems that are in a state of biaxial compression. Details of their shape and size will depend on factors such as the film stress, thickness, and interfacial toughness. The buckling geometry can be used to assess the interfacial toughness. Figure 8.12 is a picture of three straight-sided delaminations of a tungsten and DLC film stack on a silicon wafer.

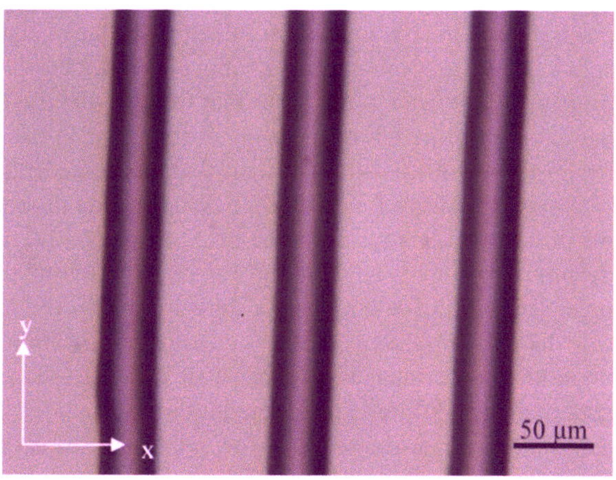

Fig. 8.12 Straight-sided delaminations of a W/DLC film on Si

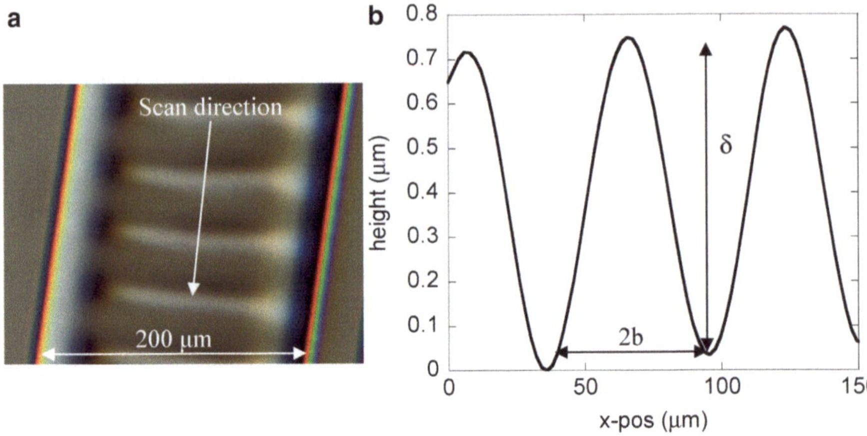

Fig. 8.13 Tungsten delamination: (**a**) Optical image of delamination morphology and (**b**) Profile of delaminations

Assuming that fracture happens at the interface, (8.2) through (8.5) can be applied to solve for the interfacial toughness between the photoresist and the tungsten film, when straight-sided delaminations are present. One of the initial deposition tests was on a silicon wafer patterned with the larger 200 μm wide photoresist lines. The tungsten film was deposited for 50 min with an argon pressure of 5 mTorr. This resulted in straight-sided delaminations forming across the width of the photoresist line, as shown in Fig. 8.13a. The profile of the straight-sided delamination widths is shown in Fig. 8.13b. Using the delamination height, δ, the delamination half width, b, and (8.2)–(8.5), the critical buckling stress was calculated to be 194 MPa. The residual stress of the film was found to be 313 MPa, the steady-state interfacial toughness was found to be 0.0114 J/m^2, and the mode-dependent interfacial toughness was found to be 0.0859 J/m^2.

The steps taken for calculating the interfacial toughness assumed that fracture took place at the tungsten/photoresist interface. There is a possibility that the crack propagates in the photoresist, at the interface between the photoresist and the silicon substrate, or a combination of above mentioned. No matter where the crack is propagating, however, the important thing is that the interfacial toughness between the tungsten film and the silicon substrate, which has been reported in literature to be 1.73 J/m^2 [30], is stronger than that measured in the photoresist areas.

Moon et al. have found that the delamination morphology can be predicted when the film stress is compared to the buckling stress [27]. For $\sigma_r/\sigma_B < 6.5$ straight-sided delaminations are predicted and for $\sigma_r/\sigma_B > 6.5$ telephone cord delaminations are predicted. For the delaminations shown in Fig. 8.13, $\sigma_r/\sigma_B = 1.6$, which agrees with the Moon et al. findings. Their predictions of delamination morphology were based on similar experimental methods that utilized lithography techniques in applying patterned areas of low interface adhesion surrounded by areas of high adhesion. By controlling the width of the low-adhesion strips, the buckle morphology was controlled.

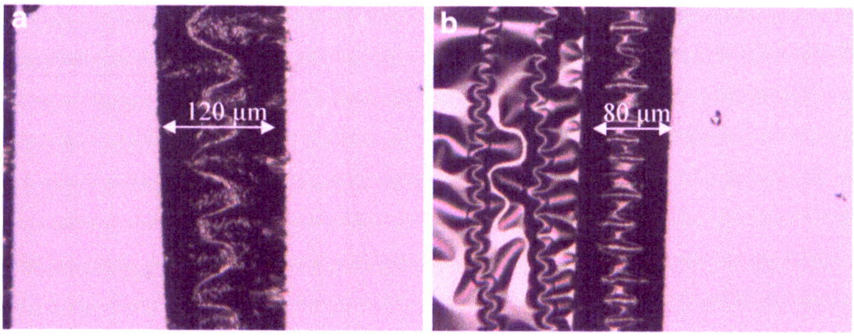

Fig. 8.14 Delamination morphology with different photoresist widths: (**a**) Telephone cord delamination and (**b**) straight-sided delamination

Ideally the delaminations would propagate parallel to the photoresist lines as opposed to perpendicular to them, as seen in Fig. 8.13. The main reason in preventing this from happening is that the photoresist line is too wide. The photoresist is so wide that it exceeds the critical buckling width, thereby allowing enough room for the straight-sided delaminations to run perpendicular to the lines. Figure 8.14 shows two different delamination morphologies that are possible when the photoresist line is decreased in width. Telephone cord delamination morphology can be observed in Fig. 8.14a when the photoresist width was approximately 120 μm. When the photoresist width was reduced to approximately 80 μm, as shown in Fig. 8.14b, a straight-sided delamination was created. Unfortunately, as shown in Fig. 8.14b, tungsten delamination also occurred in between the photoresist lines.

The delamination morphology that was observed in Fig. 8.14 would be necessary if the delaminations were to find future use in transporting fluids. Continued work is necessary to further experiment with photoresist width and its effect on delamination morphology. Attention to deposition parameters, in regard to controlling the residual stress, was found to be more important when using the adhesion-reducing layers. Not only was delamination occurring in the patterned areas, but it was also common to see the delamination propagate across other areas of the wafer. The adhesion-reducing areas acted as crack initiation sites that helped spawn delamination upon film deposition. There appears to be a fine line between creating delaminations on the patterned areas and creating them on the rest of the wafer. Controlling the exact amount of compressive residual stress is critical for this method to succeed in the consistent creation of microchannels.

8.6 Conclusions and Outlook

By using photolithography to create adhesion-reducing layers, buckling delaminations have been controlled and show potential use as microchannels. Delamination morphology depends on two conditions: (1) the buckling stress

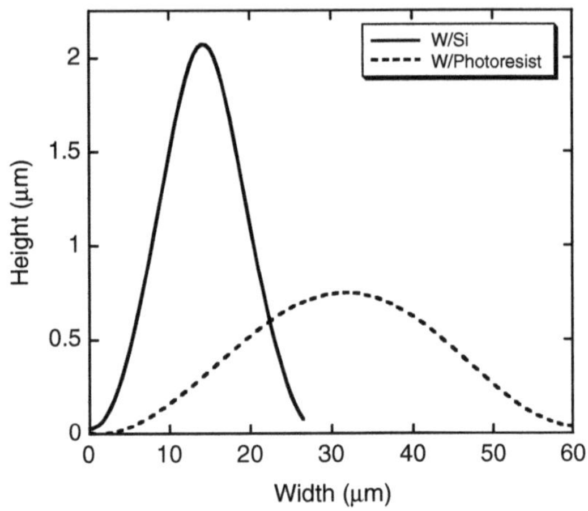

Fig. 8.15 Different delamination channel profiles

which is controlled by the adhesion-reducing layer's width and (2) the amount of compressive residual stress in the thin film. Here, telephone cord delaminations were observed at larger photoresist widths and straight-sided delaminations were observed for smaller photoresist widths. Line widths between 80 and 220 μm were created.

By using standard lithography techniques, areas of low adhesion were created to control delamination morphology. This could be used as a new method for creating microchannels for transporting, mixing, and storing fluids in microfluidic devices. Current methods for creating microchannels involve etching and wafer bonding. The potential advantages of this new method are found in its ease of manufacturing and its cost-effectiveness. Proof of concept was provided here by using a photoresist as the adhesion-reducing layer. By controlling the photoresist line width, the delamination morphology was consequently controlled.

Microfluidics has been a rapidly growing field, along with the rest of the microelectronics boom. In the late 1980s the early stages of microfluidics were dominated by the development of microflow sensors, micropumps, and microvalves [1]. Like many different areas of engineering, having everyone agree on a set definition is sometimes difficult. Microfluidics does not differ in this regard. One point that the majority can agree upon is that a microscopic quantity of fluid is the key issue in microfluidics. One main advantage of microfluidics is utilizing scaling laws for achieving better sensor performance. As the fluid volume is decreased, there will also be a need to decrease the size of the channels the fluid is transported in. Some possible delamination sizes created here are shown in Fig. 8.15. The delamination channel width was varied from 25 to 60 μm and the height was varied from 0.75 to 2 μm.

An interesting effect has been observed with the introduction of water at the film/substrate interface of highly compressed films. Water appears to reduce interfacial toughness, thereby allowing for the initiation of spontaneous delamination propagation. Figure 8.16 shows the propagation of a telephone cord delamination, over a

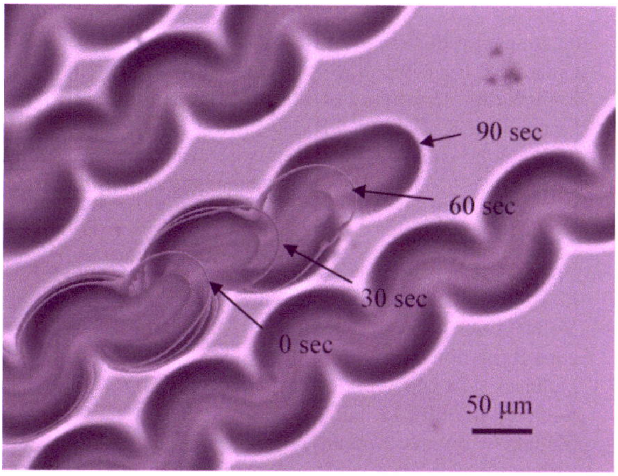

Fig. 8.16 Delamination propagation induced by the introduction of water. Waters and Volinsky [31], reprinted with permission

90-s time interval, when water was introduced at the lower left-hand corner [31]. In this case (Fig. 8.16), the delamination microchannels could be used as a one-time use, disposable microfluidic device.

Using film delaminations to create microchannels shows promise in the field of microfluidics. The key to creating the microchannels is in utilizing areas of reduced adhesion to control the delamination morphology. Possible areas for future work could be in finding better choices for adhesion-reducing layers and developing more complex delamination patterns. Creating the microchannels is only a small component of the overall picture if they are to be used in microfluidic devices. Integration of the microchannels onto a "lab-on-a-chip"-type device [32] is the overall goal, but many questions still need to be answered on how the fluid will be placed into the microchannels and how the fluid will be transported once inside the microchannels.

References

1. Nguyen NT, Werely ST (2002) Fundamentals and applications of microfluidics, Artech House, Norwood, MA,1–19:285–286
2. Spence A, Retterer S, Isaacson M (2002) Microfabricated model silicon probes with microfluidic, channels for drug delivery. NNUN Abstracts 2002/Biology & Chemistry, p 13
3. Li Y, Gulari MN, Wise KD (2003) High-yield buried microchannel formation for drug delivery at the cellular level, In: Northrup MA, Jensen KF, Harrison DJ (eds) Proceedings of mTAS 2003 seventh international conference on micro total analysis systems, vol 2, October 5–9, Squaw Valley, CA, pp 931–934
4. Volinsky AA (2003) Experiments with in-situ thin film phone cord delamination propagation. Mat Res Soc Symp Proc 749:W10.7
5. Volinsky AA, Meyer DC, Leisegang T, Paufler P (2003) Fracture patterns in thin films and multilayers. Mat Res Soc Symp Proc 795:U3.8
6. Volinsky AA, Waters P, Kiely JD, Johns EC (2005) Sub-critical telephone cord delamination propagation and adhesion measurements. Mat Res Soc Symp Proc 854E:U9.5

7. Volinsky AA, Waters P, Wright G (2004) Micro-fluidics applications of telephone cord delamination blisters. Mat Res Soc Symp Proc 855E:W3.16
8. Galambos P (1998) Two-phase dispersion in micro-channels. Ph.D. Thesis, Mechanical Engineering, University of Washington, Seattle
9. Macounova K, Cabrera CR, Holl MR, Yager P (2000) Generation of natural pH gradients in microfluidic channels for use in isoelectric focusing. Anal Chem 72:3745–3751
10. Ohring M (1992) The materials science of thin films. Academic, London
11. Stoney GG (1909) The tension of metallic films deposited by electrolysis. Proc Roy Soc Lond A82:72
12. Volinsky AA, Moody NR, Gerberich WW (2002) Interfacial toughness measurements for thin films on substrates. Acta Mater 50(3):441
13. Volinsky AA, Moody NR, Kottke ML, Gerberich WW (2002) Fiducial mark and nanocrack zone formation during thin film delamination. Philos Mag A 82:3383–3391
14. Volinsky AA, Moody NR, Gerberich WW (2003) Fiducial mark and CTOA estimates of thin film adhesion. Int J Fract 119(4):431–439
15. Matuda N, Baba S, Kinbara A (1981) Internal stress, Young's modulus and adhesion energy of carbon films on glass substrates. Thin Solid Films 81:301
16. Gille G, Rau B (1984) Buckling instability and adhesion of carbon layers. Thin Solid Films 120:109
17. Seth J, Raghunath R, Babu SV (1992) Influence of the deposition gas mixture on the structure and failure modes of diamondlike carbon films. J Vac Sci Technol A 10(2):284–289
18. Moon M-W, Jensen HM, Hutchinson JW, Oh KH, Evans AG (2002) The characterization of telephone cord buckling of compressed thin films on substrates. J Mech Phys Solids 50 (11):2355
19. Thouless MD (1990) Crack spacing in brittle films on elastic substrates. J Am Ceram Soc 73:2144
20. Bai T, Pollard DD, Gao H (2000) Explanation for fracture spacing in layered materials. Nature 403:753
21. Huang R, Prevost JH, Huang ZY, Suo Z (2003) Channel-cracking of thin films with the extended finite element method. Eng Fract Mech 70(2513)
22. Hutchinson JW, Suo Z (1992) Mixed-mode cracking in layered materials. Adv Appl Mech 29:63–191
23. Prinz VY (2000) Free-standing and overgrown InGaAs/GaAs nanotubes, nanohelices and their arrays. Physica E 6(1–4):828–831
24. Prinz VY (2004) Precise semiconductor nanotubes and nanoshells fabricated on (110) and (111) Si and GaAs. Physica E 23:260–268
25. World Wide Web: http://www.eng.usf.edu/~volinsky
26. Burolla VP (1980) Deterioration of the silver/glass interface in second surface solar mirrors. Sol Energy Mater 3/1–2:117–126
27. Moon MW, Lee KR, Oh KH, Hutchinson JW (2004) Buckle delamination on patterned substrates. Acta Mater 52/10:3151–3159
28. Volinsky AA, Vella JB, Gerberich WW (2002) Fracture toughness, adhesion and mechanical properties of low-K dielectric thin films measured by nanoindentation. Thin Solid Films 429 (1–2):201–210
29. Franssila S (2004) Introduction to micro fabrication. Wiley, West Sussex, England
30. Kriese MD, Gerberich WW, Moddy NR (1999) Quantitative adhesion measures of multilayer films: Part II. Indentation of W/Cu, W/W, Cr/W. J Mater Res 14(7):3019
31. Waters P, Volinsky AA (2006) Stress and moisture effects on thin film buckling delamination. Exp Mech 47(1):163–170
32. Weigl BH, Hedine K (2002) Lab-on-a-chip-based separation and detection technology for life science applications. Am Biotechnol Lab 20(1):28–30

Chapter 9
Mechanical Self-Assembly on Curved Substrates

Xi Chen and Jie Yin

Abstract Self-assembled buckling patterns of thin films on compliant substrates have been subjected to extensive studies and shown great promise in micro-fabrication. However, most previous studies were limited to planar substrates, and the study of buckling of films on curved substrates has not received sufficient attention. With the constraining effect from various types of substrate curvature, numerous new types of buckling morphologies may emerge which not only enable true three-dimensional (3D) fabrication of microstructures and microdevices but also have important implications for the morphogenesis of quite a few natural and biological systems. We review the scientific aspects of elastic buckling of thin films on several representative curved substrates, emphasizing the critical effect of substrate curvature, its interaction with other material/system parameters, and ways to control the buckles based on mechanical and physical principles and bridge them with prospect applications in biology, biomedical engineering, and small-scale fabrication.

X. Chen (✉)
Department of Earth and Environmental Engineering, Columbia Nanomechanics Research Center, Columbia University, New York, NY 10027, USA

SV Lab, International Center of Applied Mechanics, School of Aerospace, Xi'an Jiaotong University, Xi'an 710049, China

Department of Civil & Environmental Engineering, Hanyang University, Seoul 133-791, Korea
e-mail: xichen@columbia.edu

J. Yin
Department of Earth and Environmental Engineering, Columbia Nanomechanics Research Center, Columbia University, New York, NY 10027, USA

X. Chen (ed.), *Mechanical Self-Assembly: Science and Applications*,
DOI 10.1007/978-1-4614-4562-3_9,

9.1 Introduction

Self-assembled buckling of thin films on compliant substrates can achieve highly ordered patterns when the film deformation mismatches with that of the substrate, which has vast potential applications in stretchable interconnects [1, 2], flexible integrated circuits [3], optical gratings [4], measuring the film modulus [5], and producing wrinkled substrate to control the direction of cell growth [6], among others.

Bowden et al. [7] first reported that when a thin metal film of submicron thickness was deposited on a compliant planar PDMS substrate, spontaneous elastic buckling patterns were observed in the film as the system was cooled owing to the mismatched thermal deformation, with typical wavelengths on the order of microns. Since their pioneering experiment, extensive experimental, theoretical, and numerical studies were carried out to explore the buckling mechanisms and investigate the feasibility of quantitative control of the patterns for applications in micro- or nanostructures. Among experimental efforts, the substrate surface topology may be manipulated to change the local film stress so as to generate a variety of ordered patterns [8]. Similarly, local physical properties of the thin film can be perturbed to result in various buckle patterns [9], and more refined nanoscale patterns may be achieved by modifying the surfaces using focused ion beam [10]. External constraints may be applied, where a pre-patterned mold was held against the film as the buckles were formed, and the resulting pattern was quite stable after the removal of the mold [11]. The substrate may also be pre-strained [12], where silicon nano-ribbons bonded to a pre-stretched flat PDMS could generate wavy layouts upon releasing the substrate strain. Besides solid substrates, a recent work [13] reported wrinkle-to-fold transition of buckled thin polyester film on water or soft gel substrates.

In order to explain the formation mechanisms of various intriguing buckling patterns from theoretical aspects, Chen and Hutchinson [14, 15] showed that upon equi-biaxial compression of a film bonded to a planar semi-infinite compliant substrate, the herringbone pattern possesses less strain energy than its competitors and thus is more favorable. A more comprehensive discussion of the herringbone mode was given by Audoly and Boudaoud [16]. A study on the kinetic buckling of elastic films on viscoelastic substrates showed that both energetics and kinetics play important roles in determining the critical condition, growth rate, and wavelength [17]. Through the control of anisotropic strains in films, Huang et al. [18] simulated the evolution and transition of stripes, herringbone, and labyrinth buckling patterns. Genzer and Groenewold [19] gave extensive examples of patterns achievable via film wrinkling and bridged that with skin wrinkles and possible ways of material characterization and fabrication.

Besides the potential applications of self-assembly in micro- and nano-fabrications, spontaneous buckling instability also plays a critical role in the morphogenetic processes of some plants, including plant pattern formation, phyllotaxis, and shape of leaves. Green pioneered the biophysical explanation for the patterns commonly observed in plant shoots and flowers [20, 21], and proposed the hypothesis that buckling of the compressed tunica is the governing mechanism for determining the local phyllotactic pattern. Dumais and Steele [22] showed that since the sunflower capitulum is under circumferential compression, buckling may be a plausible explanation for the primordium initiation in the captitulum. Shipman and Newell [23] demonstrated that the local phyllotaxis and the deformation configurations on plant surfaces may be resulted from the energy-minimizing buckling pattern of a compressed shell on an elastic foundation. A recent experiment on eggplant leaf [24] showed that by applying the growth hormone auxin to the edge of the leaf to cause a large local expansion, a normally flat leaf could buckle into a wavy one. Inspired by the similarity between wrinkled edges of torn plastic sheets and a wavy leaf, Sharon et al. [25] suggested that some leaves may form wavy edges through spontaneous buckling. Recently Liang and Mahadevan [26] reproduced the typical shape of a long leaf, saddle-like in the center and wavy in the edge, by differentially stretching a foam ribbon; they proposed that the origin of the leaf morphology may arise from instability driven by the in-plane differential growth in leaves. Similar studies on *Antirrhinum* [27] and *Arabidopsis* [28] have also suggested a role for mechanical force in the regulation of local plant morphology through its interaction with biological processes.

However, most previous works were limited to the elastic buckling patterns of planar thin films (either with or without an underlying substrate that is also planar), yet the buckling characteristics of closed thin film (shell) on *curved* substrate (core) have important implications in both morphogenesis of quite a few natural and biological systems and three-dimensional self-assembly fabrication of micro- and nanostructures.

The intriguing wrinkling-like ordered patterns observed in quite a few fruits, vegetables, animal skins, tissues, and cells [29–33] may be related to mechanics-driven buckling process owing to the mismatched deformation between the shell/film (e.g., skin of fruit or membrane of cell) and the underlying curved core/substrate (e.g., flesh of fruit or cytoplasm of cell), [34] during which the curvature of the substrate plays a dominant role in shaping the distinctive overall appearance of quite a few natural and biological systems. In addition, the new varieties of buckling patterns spontaneously formed on curved substrates at the micron and submicron scales could enable true three-dimensional structures (which are otherwise difficult to fabricate using traditional techniques such as photolithography) [35, 36] and significantly expand applications in biomedical engineering [37], optics [9], optoelectronics, and display technologies [38–40].

In this chapter, we summarize some recent progresses on the mechanics and physics principles governing the elastic buckling patterns of thin films on curved compliant substrates, with a special emphasize on the effect of substrate curvature.

Consider a substrate with Young's modulus E_s and Poisson's ratio ν_s, and a thin film of thickness t, Young's modulus E_f ($E_s/E_f = 1$), and Poisson's ratio ν_f, is bonded to the substrate in the due course of buckling; both the film and substrate are assumed to be homogeneous, isotropic, and elastic unless otherwise denoted. The mismatched deformation between the film and substrate can be induced by various ways, including differential growth, thermal expansion mismatch, electric field-responsive deformation mismatch, phase transformation-induced strain mismatch, swelling or dehydration mismatch, osmotic pressure, environmental pH variation, etc., such that the overall effect is that either the substrate shrinks more than the film or the film expands more than the substrate, and that would render the film in compression and when such a stress exceeds the threshold, spontaneous buckles would occur with a distinct pattern. If the stress field in the film is anisotropic and inhomogeneous (which in this chapter is mainly caused by the effect of substrate curvature), buckles are likely to occur in the regions with more prominent stress and align in preferred directions so as to relieve the strain energy more effectively. The two most critical parameters for characterizing the buckles are the critical buckling stress and buckling wavelength.

For the extensively studied case where the substrate is planar (and semi-infinite), when the system undergoes equi-biaxial compression, very often the herringbone pattern would emerge, and the critical buckling wavelength and critical stress are [14]

$$\overline{\lambda}_{\mathrm{cr}}^{\mathrm{equi}} = 2\pi t\left(\frac{\overline{E}_{\mathrm{f}}}{3\overline{E}_{\mathrm{s}}}\right)^{\frac{1}{3}} \text{ and } \overline{\sigma}_{\mathrm{cr}}^{\mathrm{equi}} = \frac{1}{4}\left(9\overline{E}_{\mathrm{s}}{}^{2}\overline{E}_{\mathrm{f}}\right)^{\frac{1}{3}}, \tag{9.1}$$

respectively, where $\overline{E}_{\mathrm{f}} = E_{\mathrm{f}}/(1-\nu_{\mathrm{f}}^{2})$ and $\overline{E}_{\mathrm{s}} = E_{\mathrm{s}}/(1-\nu_{\mathrm{s}}^{2})$. If the compression is uniaxial the corresponding parameters are [41]

$$\overline{\lambda}_{\mathrm{cr}}^{\mathrm{uni}} = 2\pi t\Omega\left(\frac{\overline{E}_{\mathrm{f}}}{\overline{E}_{\mathrm{s}}}\right)^{\frac{1}{3}} \text{ and } \overline{\sigma}_{\mathrm{cr}}^{\mathrm{uni}} = \Xi\left(\overline{E}_{\mathrm{s}}^{2}\overline{E}_{\mathrm{f}}\right)^{\frac{1}{3}}, \tag{9.2}$$

where $\Omega = \left[(3-4\nu_{\mathrm{s}})(1-\nu_{\mathrm{s}})^{2}/12\right]^{\frac{1}{3}}$ and $\Xi = 3\left[12(3-4\nu_{\mathrm{s}})^{2}/(1-\nu_{\mathrm{s}})^{4}\right]^{-\frac{1}{3}}$.

Perhaps the easiest approach of incorporating the substrate curvature effect is to introduce a uniform curvature along one direction of the initial planar substrate, and in this case the substrate becomes cylindrical. The geometry of cylindrical-like shell/core structure is often found in micro/nanowires, fibers, plant stem, animal body, as well as some tissues such as arteries and collagen fibers. When the principal curvature is uniform in all directions, the spherical substrate, which is widely observed in bubbles, brain cortex, cells, nucleus, etc., deserves another fundamental case study. Spheroidal substrate, which may be regarded as a modification of the spherical one, is a good approximation of quite a few fruits, vegetables, eggs, etc. Finally, when the cylindrical substrate is combined with a spherical cap, a model fingertip emerges and its wrinkling pattern upon water immersion may be interesting. In what follows, for each representative type of substrate, we demonstrate the interaction and coupling between the substrate curvature effect and other material and system variables, the ways to control the characteristics of buckles, as well as the potential applications and implications of these self-assembled buckling patterns on curved substrates in morphogenesis and three-dimensional fabrications.

9.2 Cylindrical Substrate

9.2.1 *Effect of Curvature on Critical Buckling Stress and Wavelength*

Consider a long cylindrical substrate with radius R ($R/t>>1$) and length L, the lateral surface of which is completely covered by the film. In such a plane strain system with the increase of the mismatched deformation between the film and substrate, the hoop stress in the film will build up. The magnitude of the unconstrained strain mismatch between the film and substrate is $\Delta\varepsilon$. For instance, upon thermal expansion mismatch, let the coefficient of thermal expansion (CTE) of the film and substrate to be α_f and α_s, respectively, then $\Delta\varepsilon = |\alpha_f - \alpha_s|\Delta T$, where in case $\alpha_f<\alpha_s$, ΔT is the temperature drop during cooling, and in case $\alpha_f>\alpha_s$, ΔT is the temperature increase during heating. Similarly, for growth, swelling, or dehydration mismatch, one can replace α_f and α_s by the respective growth coefficients, and ΔT by the effective growth time. The magnitude of film stress in the pre-buckling state is

$$\sigma_{f0} = \frac{E_f E_s(2R^2 + 2Rt + t^2)\Delta\varepsilon}{2E_s(1 - \nu_f^2)R^2 + [E_s(1 + \nu_f) + E_f(1 + \nu_s)(1 - 2\nu_s)](2Rt + t^2)}. \tag{9.3}$$

Equation (9.3) is valid for film/shell of any thickness; for thin film, the higher order terms related to t^2 can be omitted. Figure 9.1 gives the variation of $\sigma_{f0}/\sigma_\infty$ as a function of R/t, for several different E_f/E_s values (the Poisson's ratio for both substrate and film is assumed to be 0.35 in this chapter unless otherwise denoted). Here, $\sigma_\infty = \overline{E}_f\Delta\varepsilon$ is the limit for planar substrate (i.e., when $R/t \to \infty$ in (9.3)). It can be seen that if E_f/E_s is relatively small, with the increase of the substrate radius of curvature R/t, the film stress quickly approaches to the planar limit, whereas the convergence is much slower when the film is much stiffer than the substrate; that is, the substrate curvature effect is coupled with the elastic mismatch effect.

When σ_{f0} exceeds a critical value, the film will buckle. The critical buckling wavelength can be obtained from a simplified plane-strain ring-foundation model [34], where the exact solution of the corresponding wrinkle wave number n_{cr} can be given by minimizing the following equation with respect to n [34]:

$$p_n = \frac{\overline{E}_f I(1 + \overline{K})}{R^3} \frac{\left[(n^2 - 1)^2 + \overline{K}(1 + AR^2/I)\right]}{\left[n^2 - 1 + \overline{K} - \left(\overline{K} + \overline{K}^2\right)/n^2\right]}, \tag{9.4}$$

where p_n is the critical bifurcation line pressure of the nth mode and A and I are the area and moment of inertia of the cross section of the film, respectively;

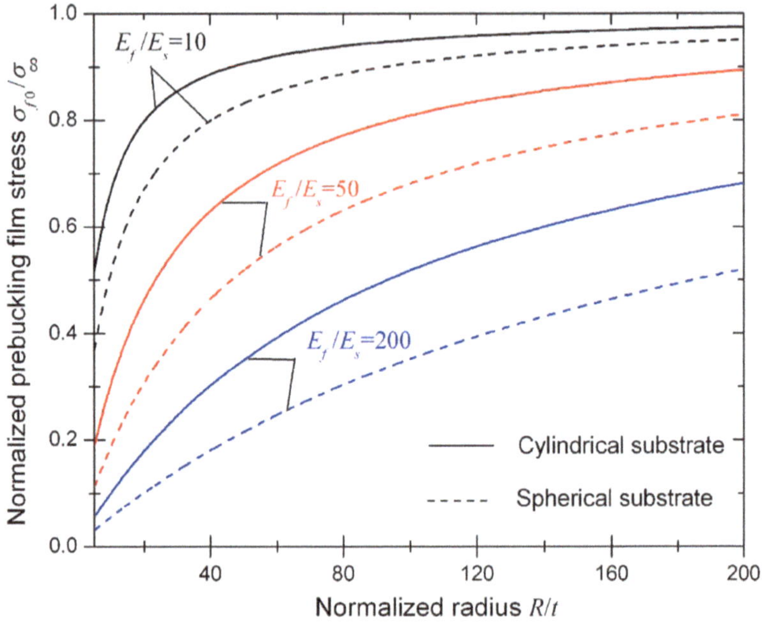

Fig. 9.1 The effect of substrate curvature (t/R) on the pre-buckling stress of thin films on cylindrical and spherical substrates with different film/substrate modulus mismatch (E_f/E_s). Here the pre-buckling stress is normalized by its counterpart for planar substrate of respective scenarios

$\overline{K} = \tilde{E}_s R / \overline{E}_f t$ is the dimensionless Winkler foundation stiffness [41] where $\tilde{E}_s = E_s/(1-2\nu_s)(1+\nu_s)$. If the substrate is inhomogeneous, e.g., containing a stiff or a compliant core, then $\overline{K}$ can be revised accordingly and (9.4) is still valid. For instance, if the substrate contains an empty core with radius a, then $\overline{K} = \frac{\overline{E}_s R}{\overline{E}_f t}\left(\frac{R^2+a^2}{R^2-a^2} - \frac{\nu_s}{1-\nu_s}\right)^{-1}$.

The exact solution of n_{cr}, however, must be derived numerically. Instead, one may simplify (9.4) by assuming that $\overline{K}$ is small compared with n, and thus the reduced solution of the critical buckling wave number n_{cr}, wavelength $\hat{\lambda}_{cr}$, and critical stress $\hat{\sigma}_{cr}$ can be derived in closed form [34]:

$$n_{cr} = \left(\frac{R}{t}\right)^{\frac{3}{4}}\left(\frac{12\tilde{E}_s}{\overline{E}_f}\right)^{\frac{1}{4}},\ \hat{\lambda}_{cr} = \frac{2\pi R}{n_{cr}} = 2\pi t\left(\frac{R}{t}\right)^{\frac{1}{4}}\left(\frac{\overline{E}_f}{12\tilde{E}_s}\right)^{\frac{1}{4}} \text{ and } \hat{\sigma}_{cr} = \left(\frac{\overline{E}_f\tilde{E}_s}{3}\right)^{\frac{1}{2}}\left(\frac{t}{R}\right)^{\frac{1}{2}}. \tag{9.5}$$

Comparing (9.5) with (9.1) and (9.2), the effect of substrate curvature t/R on buckling characteristics is obvious. Figure 9.2 shows that for a planar substrate, the critical buckling wavelength is almost the same under uniaxial and equi-biaxial compressions. For both planar and cylindrical substrates, the buckling wavelength

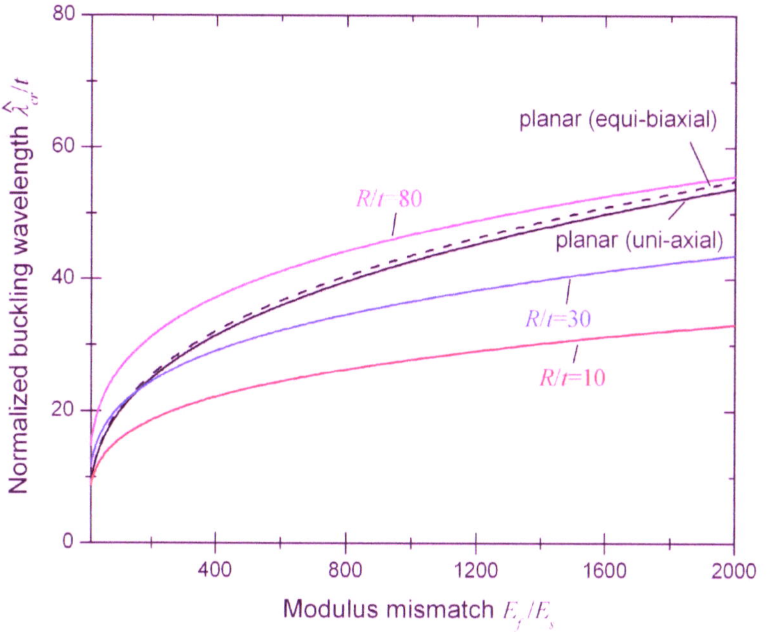

Fig. 9.2 The normalized buckling wavelength as a function of film/substrate stiffness mismatch, presented for different curvatures of the cylindrical substrate and compared with the planar counterparts under uniaxial and equi-biaxial compression. Note that since the cylindrical film is closed, as R/t is increased, its wavelength does not necessarily converge to the planar solution

increases with the modulus mismatch E_f/E_s and t. However, the wavelength of the curved substrate also depends on the normalized curvature and it increases with R/t. Moreover, for planar substrates, the critical buckling stress is independent of the film thickness, whereas t also affects the bifurcation of cylindrical film/substrate system through the term R/t. Such a qualitative comparison illustrates the importance of the substrate curvature.

The buckle amplitude A can be obtained from the deformation compatibility between the film and substrate, which is given by

$$\frac{A}{t} = \left[\frac{2}{3}\left(\frac{\sigma_f}{\hat{\sigma}_{cr}} - 1\right)\right]^{\frac{1}{2}} = \left[\frac{2}{3}\left(\frac{\varepsilon_f}{\hat{\varepsilon}_{cr}} - 1\right)\right]^{\frac{1}{2}}, \tag{9.6}$$

where σ_f and ε_f are the film stress and strain in the buckled state, respectively. $\hat{\varepsilon}_{cr} = (\tilde{E}_s t / 3\overline{E}_f R)^{\frac{1}{2}}$ is the critical buckling strain. Although the format of (9.6) is similar to that of thin film on a planar substrate [14, 42], the effect of substrate curvature is implicitly embedded in the terms ε_f and $\hat{\varepsilon}_{cr}$.

9.2.2 Application: Mechanical Self-Assembly of Gear-Like Structures

The explicit equations (9.5) and (9.6) have been validated by simulations based on finite element method (FEM) [34, 35]. They provide the basis for mechanical self-assembly of ordered patterns on cylindrical substrates, in particular gear-like profiles the geometrical properties of which (gear teeth number and teeth amplitude) may be controlled precisely via the adjustment of material property, substrate geometry, film thickness, and degree of mismatched deformation. For example, in order to increase the teeth number, one should increase R/t or decrease E_f/E_s, with the former method more effective and the latter one able to be applied if the size of gear is fixed. If one wishes to reduce the critical buckling threshold such that the gear profile emerges more easily on the cylindrical substrate, then R/t and/or E_f/E_s should be increased. To enhance the teeth amplitude, in addition to increasing $\Delta\varepsilon$, one may also reduce the critical buckling threshold by increasing R/t and/or E_f/E_s.

In ref. [35], preliminary experiment was carried where a polyvinyl chloride (PVC) film ($t = 50$ μm) was bonded on the lateral surface of a cylindrical polyurethane substrate ($R = 1$–6mm). Upon dehydration of the substrate, the gear-like buckling profile emerged the wavelength of which followed closely of that in (9.5) (and remain unchanged during the process), and the teeth amplitude increased nonlinearly with time (or the mismatch strain, (9.6)).

Figure 9.3 shows the mechanical self-assembly of gear-like profiles via spontaneous buckling of films on curved cylindrical substrates: (a) Experiment of two spur gears self-assembled on shallow cylindrical substrates (with low aspect ratio L/R), the smaller one with $R/t = 125$ and $\overline{E}_f/\overline{E}_s = 1,516$, and the larger one with $R/t = 250$ and $\overline{E}_f/\overline{E}_s = 1,273$; relevant FEM simulation is shown. (b) Experiment of bevel gears with the smaller radius $R_1/t = 150$, aspect ratio $L/R_1 = 0.4$, cone apex angle 90°, and $\overline{E}_f/\overline{E}_s = 1,273$; matching FEM simulation is shown. (c) Simulated gear formed on the external surface of a hollow cylindrical substrate [35]. (d) Simulated gear formed on the internal surface of an annular cylindrical substrate [35]. (a)–(d) are modified from ref. [35]. (e) Simulated high-aspect-ratio gear, which is similar to the gear-like profile formed on a microscale cylindrical substrate with $R/t = 1,500$ and $\overline{E}_f/\overline{E}_s = 100$ from the experiment in Ref. [8]. (f) Wrinkled surface topologies with longitudinal grooves observed on a long electrospun polymer microfiber; the inset shows the wrinkled cross-section profile [43].

In Fig. 9.3a, b, the good agreements between experiment and simulation on spur gear and bevel gear are demonstrated. The demonstrated new mechanical self-assembly technique using soft materials also has the additional advantage of biocompatibility for potential biomedical applications.

Using the same principle, wrinkles may also be created on a hollow substrate (Fig. 9.3c), or the internal surface of an annular cylindrical substrate (Fig. 9.3d), as well as fabricating a high-aspect-ratio gear (Fig. 9.3e). In the case of the high-aspect-ratio gear (Fig. 9.3e), the simulated profile is also qualitatively

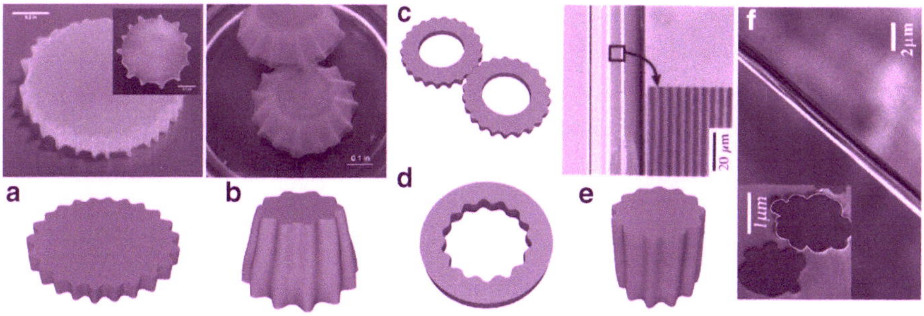

Fig. 9.3 Mechanical self-assembly of gear-like profiles via spontaneous buckling of films on curved cylindrical substrates [44], reproduced by permission of The Royal Society of Chemistry. The article in which this figure was originally published is located at the following link: http://pubs.rsc.org/en/content/articlelanding/2010/sm/c0sm00401d

consistent with an experiment by Bowden et al. [8] on a cylindrical substrate at the micron scale, where the thin film was obtained by oxidation of the surface of the elastic polymer substrate and the uniform film buckling pattern was induced by thermal expansion mismatch.

In addition to potential applications in micro-machines and soft machines [44], the gear-like wrinkled surface morphologies may largely increase the surface area of micro- or nanofibers and modify their wetting properties. For example, wrinkled surface topographies are often observed in electrospun polymer microfibers as shown in Fig. 9.3f [43]. During solvent evaporation, a glassy thin film formed on the soft cylindrical substrate, and the relative shrinkage of the substrate resulted in the wrinkled morphologies. As revealed in the surface structure of water spider legs, the microsetae with fine nanoscale axial grooves account for its remarkable water repellence [45]. Using spontaneous buckles formed on micro- and nanofibers, similar axial grooves can be engineered with tunable wavelength and amplitude (following the mechanical principles, (9.5) and (9.6)), enabling the design of superhydrophobic micro- and nanostructures.

In essence, with the underlying cylindrical substrate, the hoop stress developed in the film is about two times the axial stress, and thus the initial bifurcation always makes the wrinkles parallel to the axial direction like those in Fig. 9.3. To break the axisymmetry and create true 3D patterns on cylindrical substrates, anisotropy is needed.

9.2.3 Anisotropic Film and Implication for 3D Fabrication

If one can make the axial stress to be comparable or even higher than the hoop stress, then the circumferential wrinkles would appear. This requires anisotropy

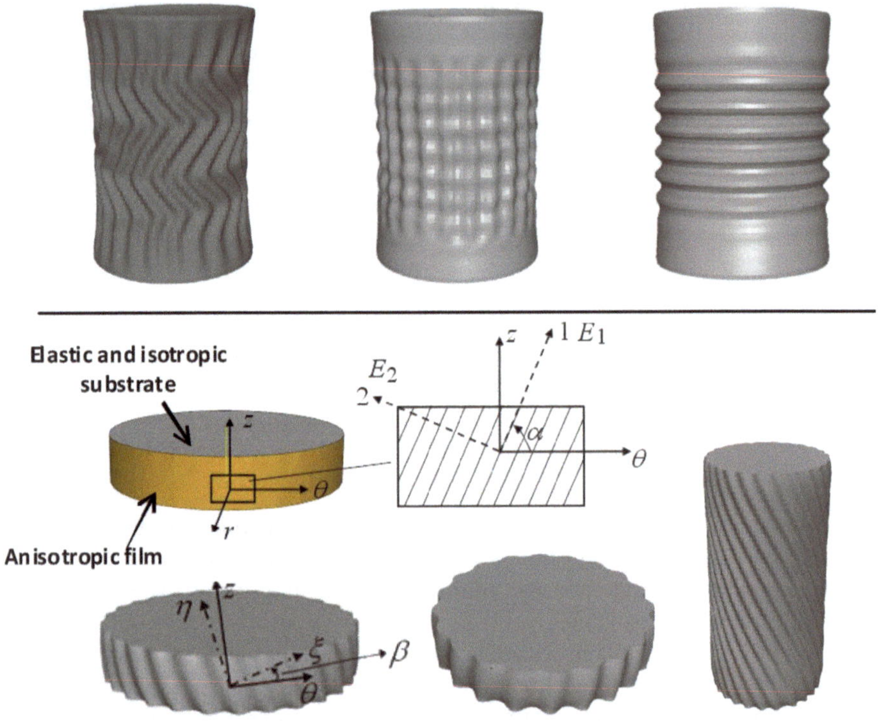

Fig. 9.4 Simulated morphologies of gears with 3D features induced by anisotropic properties. *Top row* (left to right): Effect of anisotropic growth where the mismatched strain in the axial direction is 1.5, 2, and 2.5 times that in the hoop direction. *Bottom row*: Inclined gears formed by buckling of an orthotropic film on a cylindrical substrate, where the material anisotropy axes (1, 2) are misaligned with the principal curvature axes (θ, z) of the substrate; the resulting gear pattern examples showing different teeth inclination angles and different aspect ratios [44], reproduced by permission of The Royal Society of Chemistry. The article in which this figure was originally published is located at the following link: http://pubs.rsc.org/en/content/articlelanding/2010/sm/c0sm00401d

such that either the mismatched deformation in the axial direction is much larger than that in the hoop direction, or the film stiffness in the axial direction is much smaller than that in the hoop axis. For instance, if the mismatched strain $\Delta\varepsilon$ in the axial (longitudinal) direction is 1.5, 2, or 2.5 times that in the hoop direction, numerical simulations show that the resulting hoop stress is about 1.15, 1.0, and 0.9 times the longitudinal stress; in the first row of Fig. 9.4 (with $R/t = 50, L/R = 3$, $\overline{E}_{\mathrm{f}}/\overline{E}_{\mathrm{s}} = 30$), the resulted buckle morphology becomes herringbone, square, and latitudinal, respectively, except near the edges due to the boundary effect. These new 3D profiles may find potential applications in micro-fabrication, and they may also shed some light to some natural and biological systems; for example, the latitudinal pattern is somewhat like that observed on an elephant's tail [34] or a

Shar-Pei's skin [19], indicating possible anisotropic local skin growth of the elephant or the dog.

In practice, the anisotropic film stiffness may be easier to achieve or control than anisotropic growth or shrinking. Consider an orthotropic film, the Young's modulus of which is E_1 and E_2 along the local material axes 1 and 2, is bonded to an elastic and isotropic cylindrical substrate as shown in the inset of Fig. 9.4. In general, the angle between the local film material axis and the global substrate hoop (θ) direction, α, can be nonzero. In terms of the spontaneous buckle pattern and representative examples given in the second row of Fig. 9.4, the anisotropic effect couples strongly with the curvature effect [36]. For example, the buckles do not align with the material anisotropy direction (unless the substrate is completely flat [46]), and the difference between the inclination angle of formed buckles (β) and the material anisotropy angle (α) increases nonlinearly with substrate curvature (t/R). Moreover, $\alpha - \beta$ is also largely affected by the boundary (edge) effect, and such a difference is larger when L/R is smaller [36]. In essence, β depends on α, R/t, and aspect ratio L/R, as well as the ratio between the two orthotropic moduli, $m = E_2/E_1$. If the substrate radius of curvature is relatively large ($R/t > 50$) and the edge effect is relatively small ($L/R > 4$), then β becomes close to the direction of the minimum bending stiffness of the orthotropic plate [36].

The critical buckling wavelength (i.e., the normal spacing between neighboring inclined teeth of the gear-like profile, along the ξ-direction), on the other hand, is insensitive to L/R. Along the buckled direction (ξ), the effective modulus E_ξ can be obtained via transformation of the stiffness matrix of the orthotropic film, which is a function of α, β, and orthotropic elastic constants. Based on the study of Yin and Chen [36], the critical buckling wavelength $\hat{\lambda}^{\xi}_{\mathrm{cr}}$ and critical buckling stress $\hat{\sigma}^{\xi}_{\mathrm{cr}}$ can be approximately derived as

$$\hat{\lambda}^{\xi}_{\mathrm{cr}} = 2\pi t \left(\frac{R}{t}\right)^{\frac{1}{4}} \left(\frac{E_\beta}{12\tilde{E}_{\mathrm{s}}\cos\beta}\right)^{\frac{1}{4}} \text{ and } \hat{\sigma}^{\xi}_{\mathrm{cr}} = \left(\frac{E_\beta \tilde{E}_{\mathrm{s}}\cos\beta}{3}\right)^{\frac{1}{2}} \left(\frac{t}{R}\right)^{\frac{1}{2}}. \quad (9.7)$$

Note that the form is similar to that in (9.5). Moreover, the expression of the buckle amplitude is identical to that in (9.6). Despite these similarities, we remark that in (9.7) β is also a nonlinear function of R/t; in other words, the effects of substrate curvature on the buckling characteristics (including the critical stress, wavelength, and amplitude) are much more complicated in the case of anisotropic thin film.

The simulated profiles in Fig. 9.4 are illustrative examples of true 3D patterns and microstructures that are otherwise difficult to make using conventional techniques such as photolithography (the high-aspect-ratio gear in Fig. 9.3 is another such example). It is demonstrated that mechanical self-assembly of buckles on curved substrates may provide an alternative approach to fabricate ordered 3D microstructures in a quick, simple, and cost-effective way.

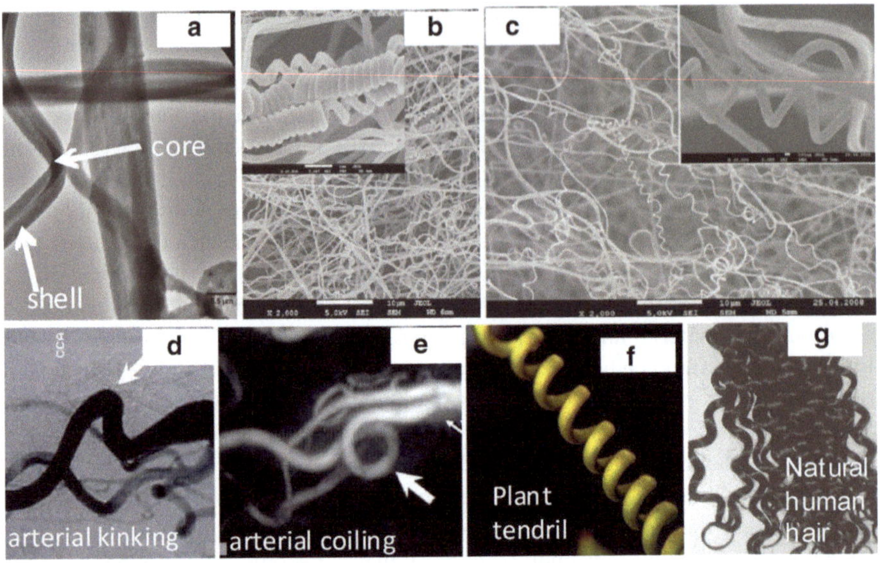

Fig. 9.5 Global instability mode of several long cylindrical shell/core structures [44], reproduced by permission of The Royal Society of Chemistry. The article in which this figure was originally published is located at the following link: http://pubs.rsc.org/en/content/articlelanding/2010/sm/c0sm00401d *For nanofibers*: (**a**) Nanofibers with stiff shell and soft core structure produced by coaxial electrospinning. With excessive shrinkage of the core, the axial compressive stress in shell triggers the global buckling of the nanofibers into (**b**) nanosprings if the core size is comparable with the shell thickness and (**c**) nanocoils if the shell is thinner. *For tissues*: (**d**) Arterial tortuosity or kinking induced by bending buckling, (**e**) coiling of human internal carotid, and similar coiling morphology are also often observed in (**f**) plant tendrils and (**g**) natural hair, among others

9.2.4 Thick Shell and Global Instability of Coiling of Long Cylindrical Shell/Core Systems

When the thickness of film (or shell) is much smaller than that of the underlying compliant substrate (or core), i.e., $t \ll R$, upon bifurcation the strain energy is mainly released through the wrinkled film surface morphology (as described in previous sections). However, when the thickness of the shell is comparable to that of the cylindrical core, under mismatched axial deformation a global instability mode may take place in long cylindrical shell/core systems, forming spring- or coil-like structures via global bending and/or twisting.

Through coaxial electrospinning, nanofibers with stiff shell and soft core structure were produced by Chen et al. [47] (Fig. 9.5a), in which, upon differential shrinkage between the shell and core, the axial compressive stress developed in the shell may trigger global buckling and form nanosprings or nanocoils, Fig. 9.5b, c. Their studies revealed that when the core radius was comparable to the shell thickness (e.g., $t/R = 0.5$ and 0.34), significant nanospring formation was observed; nanocoils may be formed with relatively large core size (e.g., $t/R = 0.21$).

In addition, the nanospring structure was also observed in soft shell/stiff core systems, suggesting comparable contributions of the strain energies in the shell and core.

Although the global bifurcation mode of coaxial long nanowires (with relatively thick shell) provides a new approach for fabricating 3D springlike and coil-like nanostructures, the underlying quantitative physical mechanisms remain unclear. Future mechanics studies are required to correlate the curvature and coil pitch of nanosprings and nanocoils with R/t, E_f/E_s, $\Delta\varepsilon$, and any geometrical defects. The solution of such a problem will also provide valuable guidance for designing nanosprings and nanocoils, which have potential applications in nanoelectromechanical system (NEMS) as sensors, transducers, resonators, and photonics [48].

Other than these inorganic nanocoils, coiling is also often observed in helices, DNA, arteries, hair, plant tendrils, etc. Arteries are subjected to significant mechanical load from the blood pressure and the longitudinal tension arising from surrounding tissue. An artery can be regarded approximately as a long circular cylinder comprising an external thick wall and a fluidic core. Under certain combined loads, mechanical buckling of arterial vessels may occur and that would lead to arterial tortuosity or kinking (Fig. 9.5d), which are associated with significant clinical complications. The similar global buckling instability mode was employed by Han [49], showing the significant effect of longitudinal strain (which may be caused by reduced axial tension or other clinical observations) on the bending buckling of arteries. In addition to the bending deformation in kinking, arterial coiling may also occur with twisting deformation, which was observed on human internal carotid arteries (Fig. 9.5e) [50]. Understanding of the global instability mode can once again shed light on the role of stress on the formation of kinking and coiling in arteries, plant tendrils (Fig. 9.5f), human/animal hair [51] (Fig. 9.5g), and many others.

9.3 Spherical Substrate

9.3.1 Effect of Sphere Curvature on Critical Buckling Stress And Wavelength

Another fundamental geometry of curved substrate is spherical, which has a uniform curvature $1/R$ in all directions. Assuming that the sphere surface is completely covered by an isotropic film, we use the same symbols to denote film and substrate properties. In such a system with the increase of the mismatched strain between the film and substrate ($\Delta\varepsilon$), the equi-biaxial stress in the film will build up. The magnitude of film stress in the pre-buckling state is

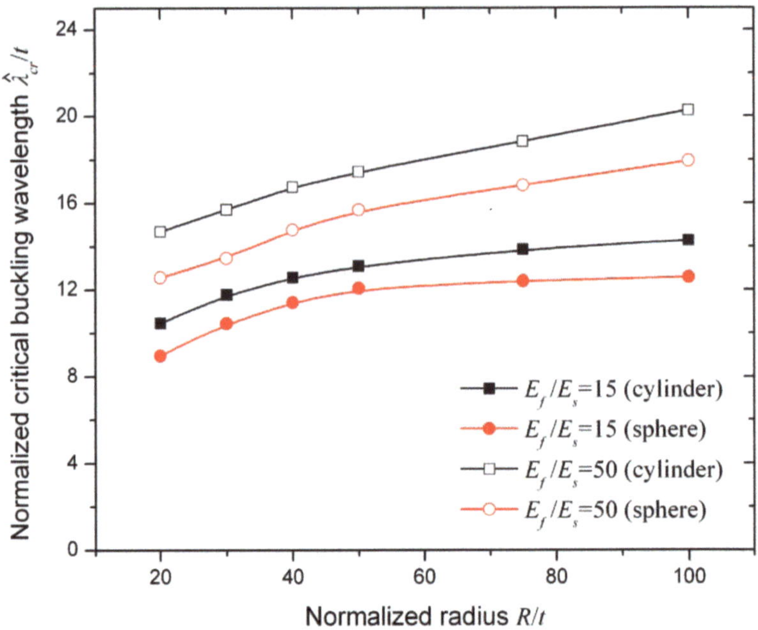

Fig. 9.6 Comparison between spherical and cylindrical substrates from FEM simulation: The normalized buckling wavelength as a function of the substrate radius of curvature. Different film/substrate stiffness mismatch ratios are shown

$$\sigma_{f0} = \frac{EE_s(3R^3 + 3R^2t + 3Rt^2 + t^3)\Delta\varepsilon}{3E_sR^3(1-v) + E_s(1+v)(3R^2t + 3Rt^2 + t^3) + 2E(1-2v_s)t(3R^2 + 3Rt + t^2)}. \tag{9.8}$$

When compared with the cylindrical counterpart in Fig. 9.1, we can see that for a spherical substrate, the film stress varies slightly slower as R/t is changed. Other trends are close to the case of cylindrical substrate. Note that (9.8) is valid for any film thickness. For thin films, the higher order terms are negligible. When σ_{f0} exceeds a critical value, the film will buckle and because the stress is isotropic and uniform, there is no preferential orientation of the buckles. When the elastic mismatch between the film and substrate is fixed, reticular and labyrinth wrinkle patterns emerge at small and large R/t, respectively (and widespread over the entire surface of the film) [52]. With increased mismatch strain, the reticular pattern may transit to labyrinth one. In Fig. 9.6 we compare the normalized buckling wavelength between spherical and cylindrical substrates (computed from FEM simulations). For both substrates, the normalized wavelength increases nonlinearly with the dimensionless substrate radius R/t and stiffness mismatch E_f/E_s. However, the wavelength in the spherical substrate is slightly lower than that in the cylindrical one when the two systems have the same curvature and modulus ratio.

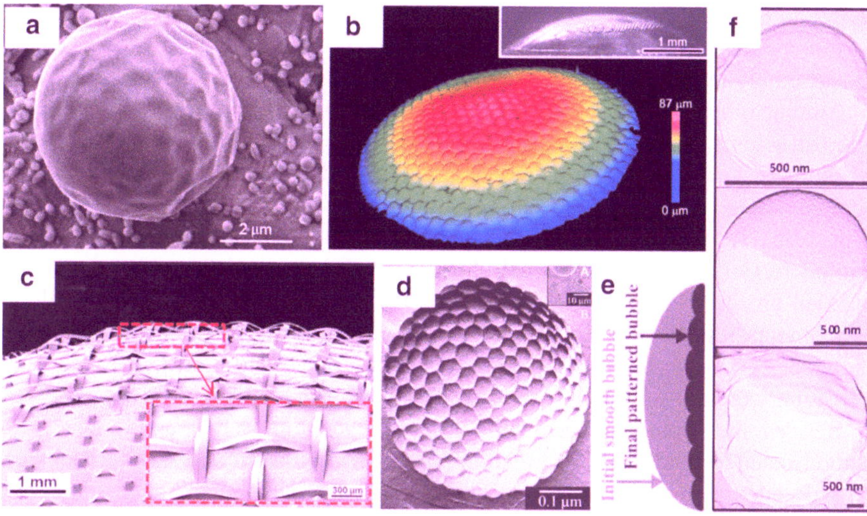

Fig. 9.7 Experiments of self-assembly on spherical shell/core systems [44], reproduced by permission of The Royal Society of Chemistry. The article in which this figure was originally published is located at the following link: http://pubs.rsc.org/en/content/articlelanding/2010/sm/c0sm00401d. *Solid inorganic systems*: (**a**) Experiment of reticular pattern formed on a spherical system (SiO_2 film/Ag substrate), with $R/t = 20$ and $\bar{E}_f/\bar{E}_s = 5$. (**b**) Microlens arrays self-assembled on a hemispherical soft substrate using constrained local buckles. (**c**) Interconnected silicon ribbonlike photodetectors on a hemispherical elastomer substrate. *Instability patterns of microbubbles*: (**d**) Nanoscale hexagonal pattern self-assembled on a stable microbubble, which is in part due to (**e**) differential shrinkage-induced buckling of the bubble surface, and (**f**) the pattern can be strongly influenced by the bubble curvature

9.3.2 Application: Mechanical Self-Assembly on Spherical Surfaces

The mechanical principles reveal that the substrate curvature t/R, elastic modulus mismatch E_f/E_s, and mismatched strain $\Delta\varepsilon$ are the governing parameters of buckling patterns. Labyrinth pattern emerges when R/t is relatively large, E_f/E_s is relatively small, or $\Delta\varepsilon$ is relatively large; the curvature effect is coupled with the other two effects. For spherical substrate, the buckling wavelength can be enhanced by increasing E_f/E_s or R/t. The critical buckling threshold can be reduced by increasing R/t or decreasing E_f/E_s, and the larger buckling amplitude can be obtained by increasing $\Delta\varepsilon$.

Cao et al. [52] demonstrated an experiment of spontaneous buckling pattern formation on spherical substrates. Using coevaporation technique, a thin SiO_2 film with thickness of about $t = 150$ nm was bonded to a nearly spherical Ag substrate at high temperature; the substrate radius varied from $R = 1$ to 50 μm. When the system was cooled owing to the thermal strain mismatch the SiO_2 film buckled, and the shape, wavelength, and critical stress conformed well to the mechanics principles. Figure 9.7a shows an example with $R = 3$ μm and at this relatively small R/t, reticular pattern was produced via spontaneous buckling. This is another example of buckling self-assembly fabrication of true 3D structures at micro- or submicron scales.

Among other limited experiments of fabricating 3D micro/nanopatterns and microstructures on curved substrates, Chan and Crosby [9] confined surface wrinkles in small local pre-patterned regions, and when such a technique was applied to a hemispherical surface, microlens arrays were self-assembled (Fig. 9.7b). Recently, Shin et al. [53] assembled interconnected silicon ribbonlike photodetectors on a hemispherical elastomer substrate, and the ribbons were in buckled profiles owing to pre-stretch (Fig. 9.7c). It should be noted that in these experiments, the buckle features created were much smaller than the substrate radius of curvature; in other words, the versatile effect and potential of substrate curvature were not fully utilized to regulate the self-assembled buckles.

Besides the aforementioned solid systems, buckling self-assembly was also demonstrated on fluidic spherical shell/hollow core microstructures, i.e., microbubbles. In a recent experiment on the nanopatterning of stable microbubbles, Dressaire et al. [54] created a nanoscale hexagonal interface pattern as shown in Fig. 9.7d through the shrinkage of the bubbles. By covering a surfactant layer on the surface of microbubbles, due to the differential shrinkage the initial smooth bubble buckled into a nano-hexagonal patterned one (Fig. 9.7e). Figure 9.7f further demonstrates the important effect of curvature on the surface wrinkling pattern, as the bubble radius was varied from 500 nm to 3 mm. Note that besides mechanical buckling, phase separation and other surface mechanisms may also underpin the pattern domains in microbubbles, and various pattern formations were reported including polygons, dendrites, beans, networks, etc. [55]. A recent review [55] has nicely summarized the nanopatterning on stable microbubbles, which have implications for biomedical applications.

It is envisioned that by coupling the intriguing substrate curvature effect with local inhomogeneous or controlled mismatch deformation, more varieties of buckling patterns that underpin micro- or nanopatterns and structures may be spontaneously fabricated on curved substrates, thereby extending the scope of the present chapter.

9.3.3 Implications for Morphogenesis of Cell and Brain Cortex

Under certain conditions, a cell with initial smooth surface morphology may become wrinkled, which is often observed in bacterial cells (Fig. 9.8a where the average wrinkle wavelength is about 100 nm) [56] and non-tissue cells such as human neutrophil cells (Fig. 9.8b) [57], macrophages, lymphocytes [58], and mast cells [59]. The wrinkled morphology may increase the surface area of the cell by more than 100% [60], which may accommodate potential membrane expansion and spreading during extravasation and osmotic swelling. From a mechanics point of view, Yin et al. [34] argued that cell wrinkling may be regarded as a buckling instability behavior induced by the mismatched deformation between the cell membrane and the cytoplasm. Such a mismatched deformation may arise from

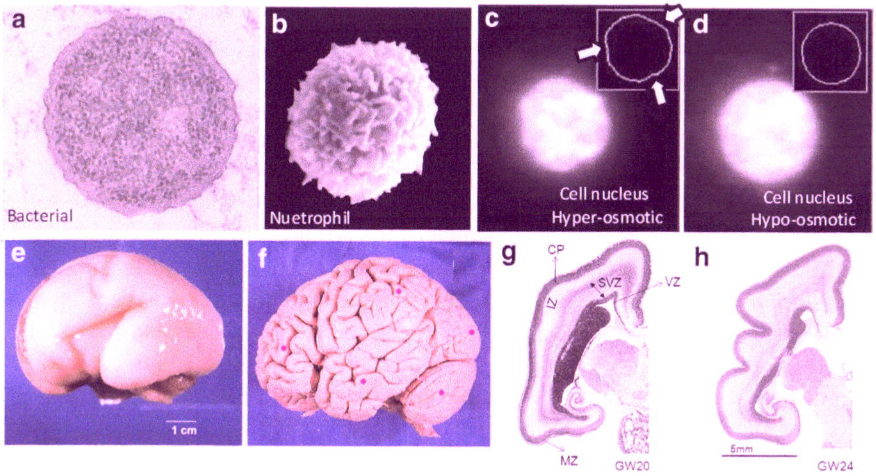

Fig. 9.8 The morphogenesis of some cells and tissues may be related to the wrinkling instability of nearly spherical shell/core systems [44], reproduced by permission of The Royal Society of Chemistry. The article in which this figure was originally published is located at the following link: http://pubs.rsc.org/en/content/articlelanding/2010/sm/c0sm00401d. *For cells*: (**a**) Wrinkled bacterial cell owing to the relative shrinkage of the cytoplasm under hyper-osmotic pressure. (**b**) Wrinkled human neutrophil cell due to the relative expansion of the cell membrane surface area during cell growth or phagocytosis. (**c**) Wrinkled cell nucleus due to hyper-osmotic shrinkage, and (**d**) the wrinkles may disappear with the swelling of nucleoplasm under hypo-osmotic pressure. *The folding pattern of brain cortex*: (**e**) The surface is relatively smooth in the fetus period yet (**f**) it folds into a pattern with bumps and grooves during growth. The cross-sectional view of brain cortex shows that (**g**) during the early stage the surface is relatively smooth, and (**f**) during the later stage the wrinkled morphology is observed

the relative shrinkage of the cytoplasm under hyper-osmotic pressure [56] (e.g., that in Fig. 9.8a) or the relative expansion of the cell membrane surface area during cell growth or phagocytosis [61] (e.g., that in Fig. 9.8b).

Other than the cell membrane surface wrinkles, recently similar wrinkled morphology was observed inside the cell, e.g., on the cell nucleus due to hyper-osmotic shrinkage, Fig. 9.8c [62]. In their work, Finan and Guilak [62] suggested that the nucleus wrinkles can be explained by the mechanical buckling of shell/core structure [34], where the contraction of the soft core (nucleoplasm) renders the stiffer shell (nuclear lamina) in compression to initiate the buckles. Under hypo-osmotic pressure, the swelling of nucleoplasm will make the lamina in tension and stretch the lamina into a smooth shape as shown in Fig. 9.8d [62].

In addition to the explanation of the wrinkle profile, the mechanical buckling principles may also enable the in situ characterization of cellular properties. For instance, within the mechanics framework the wrinkle wavelength and amplitude can be established as functions of the ratio between the shell thickness and core radius, elastic modulus ratio between the shell and core, and mismatched deformation (e.g., Fig. 9.6 for spherical shell/core systems). If by using high-resolution microscopy the wrinkle wavelength and amplitude can be measured in situ (for either cell membrane [56] or nucleus [62], e.g., Fig. 9.8), the critical cellular properties such as the effective membrane stiffness may be deduced and provide useful insight on its mechanical and biological functions [62]. It should be noted that the mechanics model discussed in this chapter is oversimplified for cells, and future refinements are needed by incorporating the osmotic pressure and other more realistic material and geometrical features of the cellular system.

Besides the cellular scale, the wrinkled morphology is also frequently observed at larger tissue or organ scales. An example of the wrinkled brain cortex is shown in Fig. 9.8f. The cortical morphology has fascinated scientists for centuries and the paradigm for the morphogenesis of folding patterns in brain cortex remains an open question. In fact, the human brain cortex is not born with wrinkles and folds; in the fetus period it is smooth (see Fig. 9.8e). As the neurons continue to divide, grow, and migrate, the cortex folds and forms a recognizable yet unique pattern of bumps and grooves. The cross section of the hemispherical cortex reveals the detailed information on the formation of the gyri (ridge) and sulci (groove) during development [63]. Figure 9.8g shows a relatively smooth cross section of the cerebral cortex at gestational week (GW) 20, where the multilayer architecture of the cerebral cortex is demonstrated, which consists of the superficial marginal zone (MZ), cortical plate (CP), intermediate zone (IZ), subventricular zone (SVZ), and ventricular zone (VZ) from exterior to interior. At GW 24, the wrinkled morphology is observed (Fig. 9.8h).

Understanding of the folding pattern of brain cortex has important implications in medical science, since that is closely associated with intelligence and some brain diseases including schizophrenia and autism [64]. Neurological disorders such as the Williams syndrome [65] and lissencephaly may result in abnormal cortical folding [66], where the folding wavelength increases and the amplitude of wrinkle becomes smaller. Recently, Zhang et al. [67] reported the reduced cortical folding in mental retardation (MR), where significantly reduced gyrification was observed in multiple brain regions compared with the healthy counterparts, which was possibly attributed to the abnormalities in the subcortical structure.

A number of hypotheses and models were proposed to explain how and why the cortex folds in a characteristic pattern from the biological, biochemical, and mechanical viewpoints [68–72]. Among them, Richman et al. [68] first proposed a mechanical buckling model where the cerebral cortex was modeled as a bilayer shell rested on a soft spherical core. The excessive growth of the shell relative to that of the core leads to the development of compressive stress in the shell, and the subsequent buckling may lead to the cortical folding. Despite some encouraging initial results, the model was oversimplified and some important features in the brain cortex development were missing, for example, the variable thickness of each layer during growth [73], the curvature effect on the folding pattern, anisotropy and heterogeneity in material properties, and the anisotropic cortical growth in the tangential and radial directions [74]. An improved shell/core model and the more sophisticated buckling mechanisms may help to understand the morphogenesis of cerebral cortex, including the healthy and the abnormal cortical folding patterns with implications to various diseases. It is envisioned that the extension of the fundamental model presented in this chapter, with the incorporation of the more realistic multilayer structure, anisotropy and heterogeneity, and growth behavior of the cerebral cortex tissues, may explain some factors affecting the fold pattern and fold number in the cerebral cortex, and provide useful insights for understanding several brain diseases.

It is arguable that the overall geometry of some cells and brain cortex may not be spherical, and they may be better modeled as spheroids. The spontaneous buckling pattern formation on spheroidal substrate is discussed next.

9.4 Spheroidal Substrate

9.4.1 Buckling Patterns and Governing Parameters

Owing to the isotropic and homogeneous film stress field, an ideal spherical substrate may lead to only two types of patterns (reticular or labyrinth); yet a simple variation of the substrate geometry to spheroidal may render more variety of 3D self-assembled buckling profiles (driven by anisotropic and inhomogeneous film stress field), which may also be bridged with the morphogenesis of quite a few natural and biological systems elucidated in the next section.

Consider a model spheroidal substrate $(x^2 + y^2)/R^2 + z^2/b^2 = 1$ in Cartesian coordinates, which is completely covered by (and remains bonded to) a thin film of thickness t. Two dimensionless factors may effectively characterize the normalized substrate radius of curvature: the normalized equator radius R/t and the aspect ratio $k = b/R$. If we neglect the relatively minor influence of the Poisson's ratio, then the buckling characteristics will additionally depend on the elastic mismatch between film and substrate, $\overline{E}_f/\overline{E}_s$, and the mismatch strain between them, $\Delta\varepsilon$. The effects of these four governing parameters on the buckling morphology are given below.

In the pre-buckling state, owing to the inhomogeneous curvature of the spheroidal system, upon mismatched deformation between film and substrate, the pre-buckling stress in the film is nonuniform (inhomogeneous) and anisotropic [34]. In a prolate spheroidal system ($k < 1.0$), for example, the hoop stress at the equator ($z = 0$) is the most prominent component and thus when the system is just above critical, longitudinal ridge-like buckles first appear near the equator. For an oblate spheroid ($k < 1.0$), circumferential undulations first occur around the polar region to release the most prominent local longitudinal stress.

With the increase of $\Delta\varepsilon$, the ridged pattern in prolate extends to the poles and the circumferential pattern in oblate extends to the equator. If $\Delta\varepsilon$ becomes sufficiently high, however, labyrinth patterns may become a more effective way of relieving strain energies, especially in those structures with large R/t and those with k closer to 1.0; in this case the labyrinth pattern will be widespread over the surface (somewhat similar to the brain cortex [68, 75]). In what follows, we focus on the effect of R/t and how it interacts with that of k and $\overline{E}_f/\overline{E}_s$, and we are more interested in the initial bifurcation mode at low $\Delta\varepsilon$ just above critical since large stress may not be preferred in biological systems nor fabrication, two of the main applications of the present chapter. The results obtained from FEM simulations are summarized in Fig. 9.9.

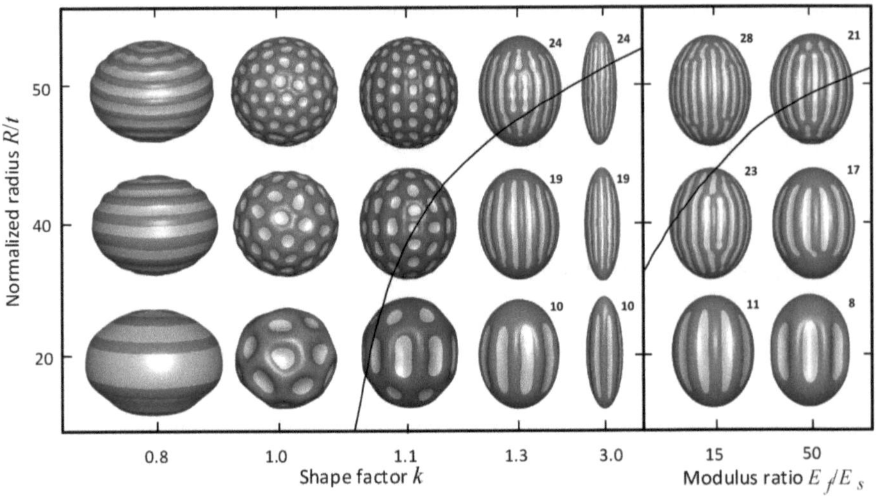

Fig. 9.9 Deformation map of spheroidal shell/core systems as the three geometrical parameters are varied [44], reproduced by permission of The Royal Society of Chemistry. The article in which this figure was originally published is located at the following link: http://pubs.rsc.org/en/content/articlelanding/2010/sm/c0sm00401d. *Left*: R/t vs. k with $\bar{E}_f/\bar{E}_s$=30. *Right*: R/t vs. $\bar{E}_f/\bar{E}_s$ with $k = 1.3$. The bright color shows the concave "bottom" of buckles. The number of longitudinal ridges is shown on *the top corner* of each relevant pattern. Here R is the equator radius and thus consistent with other geometrical profiles of curved substrate discussed in this chapter

In a distinct prolate system ($k \geq 1.3$, which is more often observed in natural and biological systems than the oblate system), when the substrate is relatively small with low value of R/t, the ridged pattern prevails and the wave number of ribs increases rapidly with R/t. Meanwhile, when R/t is large the formation of circumferential waves is possible, and thus the reticular buckles may appear almost uniformly on the surface (especially when k is small, close to the case of spherical substrate [52]). The interaction between R/t and $\overline{E}_f/\overline{E}_s$ further shows that at small $\overline{E}_f/\overline{E}_s$ and large R/t, reticular pattern has advantage over ribbed one, and with the increase of modulus mismatch, the number of ribs in the undulated film decreases. When the substrate curvature is relatively large ($R/t < 40$), the ribbed patterns may remain stable for a large range of modulus ratio ($5 \leq \overline{E}_f/\overline{E}_s \leq 200$) although the rib number may vary.

Quantitatively, for distinct prolates, the ridge number is a function of R/t and $\overline{E}_f/\overline{E}_s$, which can be well described by (9.5); similarly, (9.6) works fairly well for predicting the buckle amplitude of prolate systems. Although these two equations should, in principle, be applied to cases where k approaches infinity, extensive analyses show that the principles work quite well in general for $k \geq 1.3$ [34]. Therefore, the ways to control the buckle wave number and amplitude for distinct prolate film–substrate systems are essentially the same as those described in Sect. 9.2.2, which underpin the application below.

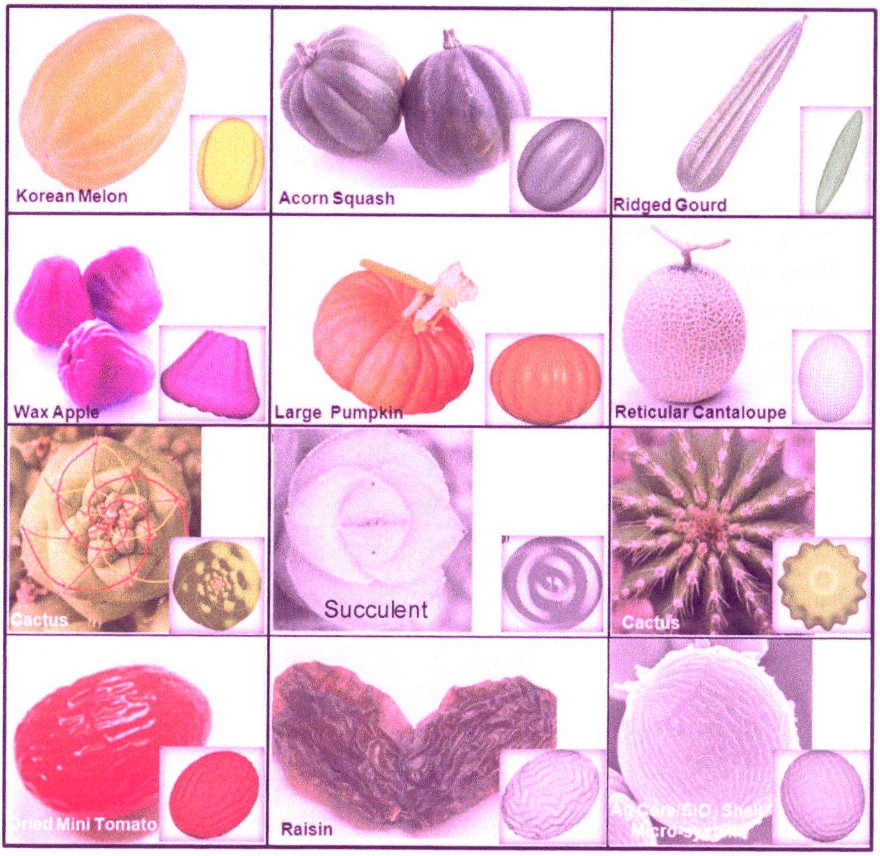

Fig. 9.10 Morphogenesis of spheroidal-like natural and biological systems, where in each example, the observation (*larger picture*) matches reasonably with the simulation based on the simple spheroidal shell/core model (*smaller picture*). [44], reproduced by permission of The Royal Society of Chemistry. The article in which this figure was originally published is located at the following link: http://pubs.rsc.org/en/content/articlelanding/2010/sm/c0sm00401d

9.4.2 Application to Organic and Inorganic Systems: Morphogenesis and 3D Fabrication

Many fruits, including varieties of cucumis melons, gourds, tomatoes, peppers, and pumpkins, take approximate spheroidal shapes with stiff skin (film/shell) bonded to compliant flesh (substrate/core) and exhibit distinctive wrinkle-like undulation morphologies (Fig. 9.10). For example, the Korean melon and ridged gourd are distinguishable by 10 equidistant longitudinal ridges. Small pumpkins, acorn squashes, and carnival squashes often have about 10 uniformly spaced ribs, whereas the large pumpkins often have about 20 or more ridges. Cantaloupes exhibit reticular patterns on their surfaces. More complicated and intriguing phyllotactic patterns are often observed in plant shoots, flowers, and the cactus head. Such distinctive yet elementary global (overall) features may be contributed in part by the stress-driven spontaneous buckling (due to the mismatched growth between the stiff skin and compliant flesh), the bifurcation mode of which may set up the template for parallel biological processes [29, 34].

Using the simple spheroidal shell/core model, various global appearances of quite a few fruits and vegetables can be reproduced [29, 34]; several examples obtained from FEM simulations are shown in the first two rows in Fig. 9.10, where a distinct morphology emerges when the governing parameters (k, R/t, $\overline{E}_f/\overline{E}_s$) are within a particular range. In the first and second row of Fig. 9.10, the morphologies of several fruits and vegetables, including Korean melon, acorn squash, ridged gourd, wax apple, large pumpkin, and cantaloupe, can be reproduced via the spheroidal shell/core model. In the third row of Fig. 9.10, plant phyllotaxis including cactus and succulent (1st and 2nd cases) can be explained using instability theory, from ref. [23]; the model of spheroidal elastic shell/foundation model may also explain the ridged pattern in cactus (3rd case). In the last row of Fig. 9.10, the same principles are applicable to dehydrated fruits such as dried mini tomato and raisin, as well as in 3D self-assembly application using SiO_2 film/Ag substrate. Moreover, the stable ribbed patterns are insensitive to minor perturbations, such as the inclusion of a rigid or a compliant core, or boundary constraints at poles [34]. Major perturbations, including the variation of the substrate curvature, anisotropic growth rate, or elastic properties, do affect the buckling pattern considerably [34], which echoes those described earlier. The same principles can be easily extended to those observed on eggs, dehydrated fruits (last row of Fig. 9.10), cells, tissues, etc. [29, 34]; some have been discussed earlier for spherical substrates and are extendable to spheroidal systems.

Besides these global morphologies, the local phyllotaxis on plant surfaces may also be due to similar buckling instability. Shipman and Newell [23] reproduced a wide spectrum of plant patterns observed in nature, and two examples of cactus and succulent are given in the third row of Fig. 9.10. The cactus can also be modeled as a thin spheroidal elastic shell rested on an elastic foundation, and such a model may explain the configuration transition between the whorl pattern and the ridged pattern observed in cactus (third image in the third row of Fig. 9.10).

Regarding engineering applications, for the same 3D fabrication experiment carried out in Sect. 9.3.2, when the inorganic substrate is spheroidal-like, the last image of Fig. 9.10 shows the SEM photo of the rib-dominated buckled patterns observed in a prolate system (Ag core/SiO_2 shell), where $k = 1.15$, $R/t \approx 55$, and $\overline{E}_f/\overline{E}_s \approx 5$, respectively. In all examples demonstrated in this section, the profiles obtained from numerical simulation, the buckling shape and wavelength predicted from theoretical equations (e.g., (9.5) and (9.6)), and the experimental (or practical) observations agree well with each other.

9.5 Combined Cylindrical and Spherical Substrates

Most previous mechanics studies on skin wrinkles were confined to planar substrates [19] and there was little emphasis on the potential influence of the underlying substrate tissue curvature. We use the common phenomenon that everybody has experienced, the wrinkles appear on human fingertip upon water immersion, to illustrate the role of substrate curvature and further bridge the buckling mechanism with morphology of biological components. From recent advances in physiological studies [76–78], the driving force for fingertip skin wrinkle is believed due to vasoconstriction of the underlying tissue, which decreases the turgor and effectively shrinks the tissue volume to produce compressive stress in the skin [79].

As a first-order approximation, the substrate of fingertip may be regarded as roughly half cylinder topped with a hemisphere (the flat surface on the backside corresponds to the "nail" region that does not wrinkle). A typical radius of the substrate is $R = 7.5$ mm and skin thickness is $t = 0.15$ mm. When the compressive stress in the film is just above critical, longitudinal wrinkles first appear near the conjunction of the cylindrical and spherical parts (which corresponds to the center of finger pad). With continued increase of $\Delta\varepsilon$ (i.e., with prolonged water immersion time), the longitudinal wrinkles propagate along the cylindrical pad whereas reticular concaves are observed on the spherical fingertip. When $\Delta\varepsilon$ is relatively large, the labyrinth pattern takes over on the finger pad, Fig. 9.11 [79]. The wrinkle amplitude increases nonlinearly with $\Delta\varepsilon$ whereas the wavelength remains almost a constant. Owing to the boundary constraint, along the circumferential direction the wrinkle amplitude is the largest at the center of finger pad and decays away near the nail region. Since the wrinkles in the cylindrical section are more prominent, the mechanical principles developed in Sect. 9.2 apply well; for example if the substrate curvature is increased (while other parameters remain fixed, e.g., from thumb to little finger), the wrinkle wavelength and amplitude decrease, and the wrinkles become more difficult. All these features are qualitatively consistent with practical observations.

The substrate geometry can be further refined based on the anatomical structure of the human fingertip; moreover, in order to capture the skin behavior more realistically, the multilayer structure of the skin and tissue can be taken into account, which includes stratum corneum, viable epidermis, dermis, subcutaneous, and bone [79]. The simulation example shown in Fig. 9.11c reveals similar features as those using the simplified model (Figs. 9.11a, b). A parametric study can be carried out to explore the role of each individual layer on wrinkling characteristics, which provide useful insights into skin aging and suppressing wrinkles [79], hence useful for biomedical and cosmetic science.

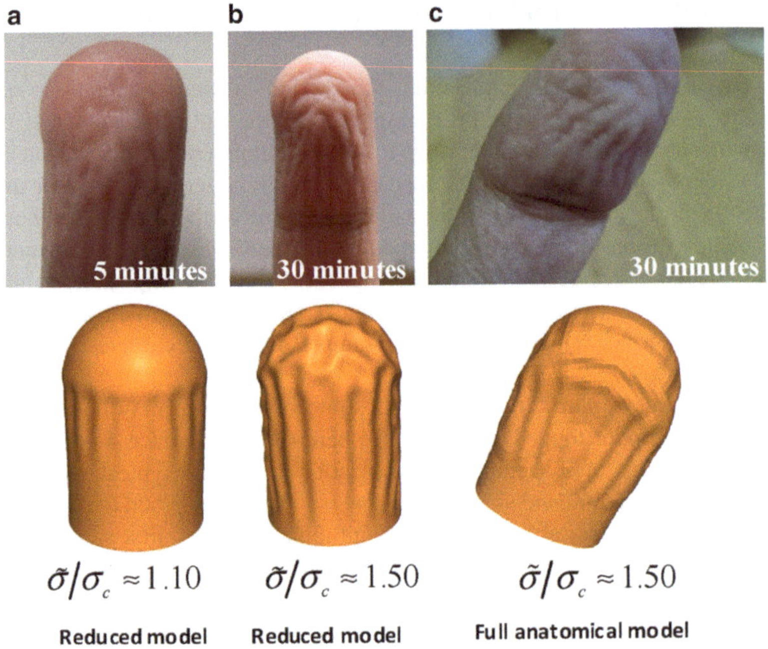

Fig. 9.11 Wrinkling of water-immersed fingertips. *First row*: Representative wrinkled morphology of fingertips (*index and thumb*); *second row*: simulations from the corresponding reduced model and full anatomical model with qualitative agreement. [44], reproduced by permission of The Royal Society of Chemistry. The article in which this figure was originally published is located at the following link: http://pubs.rsc.org/en/content/articlelanding/2010/sm/c0sm00401d

9.6 Concluding Remarks

The main objective of this chapter is to summarize some recent research progresses of the buckling patterns self-assembled in thin films on curved compliant substrates, including the underlying mechanical principles, the ways to control them, and relevant applications, which may stimulate future works on controlling the buckle morphologies by efficiently using the substrate curvature effect.

On a curved substrate depending on its principal curvature, upon mismatched deformation between the film and substrate, the compressive stress generated in the film may be anisotropic and nonuniform. This enables various types and features of buckling patterns that underpin mechanical self-assembly fabrication of 3D microstructures and devices, as well as being potentially linked to various intriguing morphological problems in biology and biomedical engineering. Figure 9.12 summarizes the possible interactions among mechanics, morphogenesis, and fabrication, all based on the wrinkling patterns of films on curved substrates. Although the central role of mechanical principle is emphasized (Fig. 9.12), we note that in terms of morphogenesis, the hypothesis of the integrated effort of the mechanical and biological processes needs to be verified from future

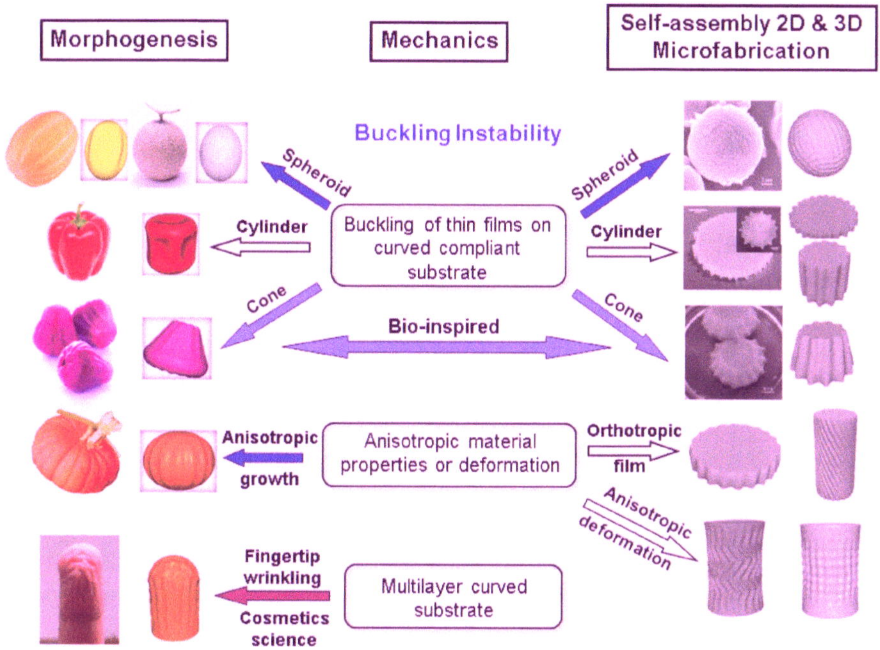

Fig. 9.12 Postulated interactions among mechanics, morphogenesis, and fabrication. *Mechanics* vs. *morphogenesis*: Due to certain biological processes occurring at the cellular level in some systems [22, 83] the film and substrate undergo mismatched deformation, and the resulting stress-driven buckling sets up a template for the overall system morphology. The processes of cell growth and cell differentiation [21, 84–87] may follow the mechanical template to help stabilize the global pattern features. On the other hand, the bifurcation mode is also closely related to the effective material and geometrical properties (which are based on biological processes). *Mechanics* vs. *fabrication*: Mechanical buckling further underpins the novel technique of self-assembly fabrication of 3D microstructures, which is inherently simple, quick, and cost-effective

biological experiments; in terms of micro- or nanofabrication, the effectiveness of the new technique requires further validation via small-scale experiments (which can be guided by the underlying mechanical principles), and additional challenges associated with defects need to be thoroughly investigated.

This chapter focuses on a few representative substrate shapes. More inspirations may be obtained from the unlimited possible types of substrate geometries in natural and biological systems and engineering practices. Moreover, the film and substrate properties can be anisotropic (as illustrated in Sect. 9.2.3) [36], nonuniform and heterogeneous, of gradient, elastoplastic or viscoelastic, and the system may be subject to external load, gravity, pre-strain, etc.—some of these scenarios were already demonstrated on planar film/substrate systems [12, 46, 80, 81] and they may lead to more varieties of spontaneous buckling patterns on curved substrates. Furthermore, the buckles may be constrained in local regions [9, 53] where the boundary effect can further interact with the substrate curvature effect. Buckle delamination pattern on curved substrate [82] is another broad and open area. For biological systems refined and more realistic models of cell/tissue growth and response are required. These will be examples of interesting problems to be explored in future, from both fundamental (underlying mechanical/physical principles) and application (soft matter in biology, biomedical engineering, and small-scale fabrication) point of views and across different length scales.

References

1. Lacour S, Wagner S, Huang Z, Suo Z (2003) Stretchable gold conductors on elastomeric substrates. Appl Phys Lett 82:2404
2. Watanabe M, Shirai H, Hirai T (2002) Wrinkled polypyrrole electrode for electroactive polymer actuators. J Appl Phys 92:4631
3. Kim D-H, Ahn J-H, Choi WM, Kim H-S, Kim T-H, Song J, Huang YY, Liu Z, Lu C, Rogers JA (2008) Stretchable and foldable silicon integrated circuits. Science 320:507
4. Harrison C, Stafford CM, Zhang W, Karim A (2004) Sinusoidal phase grating created by a tunably buckled surface. Appl Phys Lett 85:4016
5. Stafford CM, Harrison C, Beers KL, Karim A, Amis EJ, Vanlandingham MR, Kim HC, Volksen W, Miller RD, Simonyi EE (2004) A buckling-based metrology for measuring the elastic moduli of polymeric thin films. Nature Mater 3:545
6. Fleming RG, Murphy CJ, Abrams GA, Goodman SL, Nealey PF (1999) Effect of synthetic micro- and nanostructured surfaces on cell behavior. Biomaterials 20:573
7. Bowden N, Brittain S, Evans AG, Hutchinson JW, Whitesides GM (1998) Spontaneous formation of ordered structures in thin films of metals supported on an elastomeric polymer. Nature 393:146
8. Bowden N, Huck WTS, Paul KE, Whitesides GM (1999) The controlled formation of ordered, sinusoidal structures by plasma oxidation of an elastomeric polymer. Appl Phys Lett 75:2557
9. Chan EP, Crosby AJ (2006) Fabricating Microlens Arrays by Surface Wrinkling. Adv Mater 18:3238
10. Moon M-W, Lee SH, Sun J-Y, Oh KH, Vaziri A, Hutchinson JW (2007) Controlled formation of nanoscale wrinkling patterns on polymers using focused ion beam. Scr Mater 57:747
11. Yoo PJ, Suh KY, Park SY, Lee HH (2002) Physical self-assembly of microstructures by anisotropic buckling. Adv Mater 14:1383
12. Jiang HQ, Khang D-Y, Song ZJ, Sun YG, Huang YG, Rogers JA (2007) Finite deformation mechanics in buckled thin films on compliant supports. Proc Natl Acad Sci USA 104:15607
13. Pocivavsek L, Dellsy R, Kern A, Johnson S, Lin B, Lee KYC, Cerda E (2008) Stress and fold localization in thin elastic membranes. Science 320:912
14. Chen X, Hutchinson JW (2004) Herringbone buckling patterns of compressed thin films on compliant substrates. J Appl Mech 71:597
15. Chen X, Hutchinson JW (2004) A family of herringbone patterns in thin films. Scr Mater 50:797
16. Audoly B, Boudaoud A (2008) Buckling of a stiff film bound to a compliant substrate—Part II: A global scenario for the formation of herringbone pattern. J Mech Phys Solids 56:2422
17. Huang R (2005) Kinetic wrinkling of an elastic film on a viscoelastic substrate. J Mech Phys Solids 53:63
18. Huang ZY, Hong W, Suo Z (2005) Nonlinear analyses of wrinkles in films on soft elastic substrates. J Mech Phys Solids 53:2101
19. Genzer J, Groenewold J (2006) Soft matter with hard skin: From skin wrinkles to templating and material characterization. Soft Matter 2:310
20. Green PB, Steele CS, Rennich SC (1996) Phyllotactic patterns: a biophysical mechanicsm for their origin. Ann Bot 77:512

21. Green PB (1999) Expression of pattern in plants: combining molecular and calculus-based biophysical paradigms. Am J Bot 86:1059
22. Dumais J, Steele CR (2000) New evidence for the role of mechanical forces in the shoot apical meristem. J Plant Growth Regul 19:7
23. Shipman PD, Newell AC (2004) Phyllotactic patterns on plants. Phys Rev Lett 92:168102
24. Sharon E, Marder M, Swinney HL (2004) Leaves, flowers and garbage bags: Making waves. Am Sci 92:254
25. Sharon E, Roman B, Swinney HL (2007) Geometrically driven wrinkling observed in free plastic sheets and leaves. Phys Rev E 75:046211
26. Liang H, Mahadevan L (2009) The shape of a long leaf. Proc Natl Acad Sci USA 106:22049
27. Nath U, Crawford BCW, Carpenter R, Coen E (2003) Genetic control of surface curvature. Science 299:1404
28. Fleming AJ, McQueen-Mason S, Mandel T, Kuhlemeier C (1997) Induction of leaf primordia by the cell wall protein expansin. Science 276:1415
29. Yin J, Cao Z, Li C, Sheinman I, Chen X (2008) Stress-driven buckling patterns in spheroidal core/shell structures. Proc Natl Acad Sci USA 105:19132
30. Kuhl E, Maas R, Himpel G, Menzel A (2007) Computational modeling of arterial wall growth. Biomechan Model Mechanobiol 6:321
31. Hallett MB, von Ruhland CJ, Dewitt S (2008) Chemotaxis and the cell surface-area problem. Nature Rev Mol Cell Biol 9:662
32. Wang B, Liu P, Jiang W, Pan H, Xu X, Tang R (2008) Yeast cells with an artificial mineral shell: Protection and modification of living cells by biomimetic mineralization. Angew Chem Int Ed 47:3560
33. Rahmy TR, Ayoub MA (2002) In vitro action of cobra venom on goat spermatozoa ultrastructure by transmission and scanning electron microscopy. J Venomous Animals Toxins 8:127
34. Yin J, Chen X, Sheinman I (2009) Anisotropic buckling patterns in spheroidal film/substrate systems and their implications in some natural and biological systems. J Mech Phys Solids 57:1470
35. Yin J, Bar-Kochba E, Chen X (2009) Mechanical self-assembly fabrication of gears. Soft Matter 5:3469
36. Yin J, Chen X (2010) Buckling of anisotropic films on cylindrical substrates: insights for self-assembly fabrication of 3D helical gears. J Phys D: Appl Phys 43:115402
37. Jiang X, Bruzewicz DA, Wong AP, Piel M, Whitesides GM (2005) Directing cell migration with asymmetric micropatterns. Proc Natl Acad Sci USA 102:975
38. Paul KE, Prentiss M, Whitesides GM (2003) Patterning spherical surfaces at the two-hundred-nanometer scale using soft lithography. Adv Funct Mater 13:259
39. Jin H, Abelson JR, Erhardt MK, Nuzzo RG (2004) Soft lithographic fabrication of an image sensor array on a curved substrate. J Vac Sci Technol B 22:2548
40. Ruchhoeft P, Colburn M, Choi B, Nounu H, Johnson S, Bailey T, Damle S, Stewart M, Ekerdt J, Sreenivasan SV, Wolfe JC, Willson CG (1999) Patterning curved surfaces: template generation by ion beam proximity lithography and relief transfer by step and flash imprint lithography. J Vac Sci Technol B 17:2965
41. Allen HG (1969) Analysis and design of structural sandwich panels. Pergamon, New York
42. Huang Z, Hong W, Suo Z (2004) Evolution of wrinkles in hard films on soft substrates. Phys Rev E 70:030601
43. Wang L, Pai C, Boyce MC, Rutledge GC (2009) Wrinkled surface topographies of electrospun polymer fibers. Appl Phys Lett 94:151916
44. Chen X, Yin J (2010) Buckling patterns of thin films on curved compliant substrates with applications to morphogenesis and three-dimensional micro-fabrication. Soft Matter 6:5667
45. Gao X, Jiang L (2004) Water-repellent legs of water striders. Nature 432:36
46. Im SH, Huang R (2008) Wrinkle patterns of anisotropic crystal films on viscoelastic substrates. J Mech Phys Solids 56:3315

47. Chen S, Hou H, Hu P, Wendorff JH, Greiner A, Agarwal S (2009) Polymeric nanosprings by bicomponent electrospinning. Macromol Mater Eng 294:265
48. Gao PX, Ding Y, Mai W, Hughes WL, Lao C, Wang ZL (2005) Conversion of zinc oxide nanobelts into superlattice-structured nanohelices. Science 309:1700
49. Han H (2007) A biomechanical model of artery buckling. J Biomech 40:3672
50. Barbera GL, Marca GL, Martino A, Verde RL, Valentino F, Lipari D, Peri G, Cappello F, Valentino B (2006) Kinking, coiling, and tortuosity of extracranial internal carotid artery: is it the effect of a metaplasia? Surg Radiol Anat 28:573
51. Xu B, Chen X (2010) The role of stress on the formation and selection of human hair curly pattern. J Mech Behav Biomed Mater (in press)
52. Cao G, Chen X, Li C, Ji A, Cao Z (2008) Self-assembled triangular and labyrinth buckling patterns of thin films on spherical substrates. Phys Rev Lett 100:036102
53. Shin G, Jung I, Malyarchuk V, Song J, Wang S, Ko HC, Huang Y, Ha JS, Rogers JA (2010) Micromechanics and advanced designs for curved photodetector arrays in hemispherical electronic-eye cameras. Small 6:851
54. Dressaire E, Bee R, Bell DC, Lips A, Stone HA (2008) Interfacial polygonal nanopatterning of stable microbubbles. Science 320:1998
55. Borden M (2009) Nanostructural features on stable microbubbles. Soft Matter 5:716
56. Schwarz H, Koch AL (1995) Phase and electron microscopic observations of osmotically induced wrinkling and the role of endocytotic vesicles in the plasmolysis of the Gram-negative cell wall. Microbiology 141:3161
57. Bessis M (1973) Living blood cells and their ultrastructure. Springer, Berlin
58. Tohya K, Kimura M (1998) Ultrastructural evidence of distinctive behavior of L-selectin and LFA-1 alpha (L) beta (2) integrin on lymphocytes adhering to the endothelial surface of high endothelial venules in peripheral lymph nodes. Histochem Cell Biol 110:407
59. Burwen SJ, Satir BH (1977) Plasma membrane folds on the mast cell surface and their relationship to sectretory activity. J Cell Biol 74:690
60. Dewitt S, Hallett M (2007) Leukocyte membrane "expansion": a central mechanism for leukocyte extravasation. J Leukoc Biol 81:1160
61. Hallett MB, Dewitt S (2007) Ironing out the wrinkles of neutrophil phagocytosis. Trends Cell Biol 17:209
62. Finan JD, Guilak F (2010) The effects of osmotic stress on the structure and function of the cell nucleus. J Cell Biochem 109:460
63. Bayer S, Altman J (2005) The human brain during the second trimester. Taylor & Francis, Boca Raton
64. Levitt JG, Blanton RE, Smalley S, Thompson PM, Guthrie D, McCracken JT, Sadoun T, Heinichen L, Toga AW (2003) Cortical sulcal maps in autism. Cereb Cortex 13:728
65. Essen DCV, Dierker D, Snyder AZ, Raichle ME, Reiss AL, Korenber J (2006) Symmetry of cortical folding abnormalities in Williams syndrome revealed by surface based analyses. J Neurosci 26:5470
66. Levin D, Barnes PD (1999) Cortical maturation in normall and abnormal fetuses as assessed with prenatal MR imaging. Radiology 210:751
67. Zhang Y, Zhou Y, Yu C, Lin L, Li C, Jiang T (2010) Reduced cortical folding in mental retardation. Am J Neuroradiol 31:1063
68. Richman DP, Stewart RM, Hutchinson JW, Caviness VSJ (1975) Mechanical model of brain convolutional development. Science 189:18
69. Essen DCV (1997) A tension-based theory of morphogenesis and compact wiring in the central nervous system. Nature 385:313
70. Armstrong CL, Hawkes R (2000) Pattern formation in the cerebellar cortex. Biochem Cell Biol 78:551
71. Geng G, Johston LA, Yan E, Britto JM, Smith DW, Walker DW, Egan GF (2009) Biomechanisms for modelling cerebral cortical folding. Med Image Anal 13:920

72. Lefevre J, Mangin J (2010) A Reaction–diffusion model of human brain development. PLoS Comput Biol 6:e1000749
73. Kriegstein A, Noctor S, Martinez-Cerdeno V (2006) Patterns of neural stem and progenitor cell division may underlie evolutionary cortical expansion. Nat Rev Neurosci 7:883
74. Vaccarino FM, Fagel DM, Ganat Y, Maragnoli ME, Ment LR, Ohkubo Y, Schwartz ML, Sibereis J, Smith KM (2007) Astroglial cells in development, regeneration, and repair. Neuroscientist 13:173
75. Toro R, Perron M, Pike B, Richer L, Veillette S, Pausova Z, Paus T (2008) Brain size and folding of the human cerebral cortex. Cereb Cortex 18:2352
76. Wilder-Smith EPV, Chow A (2003) Water-immersion wrinkling is due to vasoconstriction. Muscle Nerve 27:307
77. Wilder-Smith EPV (2004) Water immersion wrinkling, physiology and use as an indicator of sympathetic function. Clin Auton Res 14:125
78. Hsieh C, Huang K, LiLiang P, Jeng S, Tsai H (2006) Paradoxical response to water immersion in replanted fingers. Clin Auton Res 16:223
79. Yin J, Gerling G, Chen X (2010) Mechanical modeling of wrinkled fingertip immersed in water. Acta Biomater 6:1487
80. Colin J, Coupeau C, Grilhé J (2007) Plastic folding of buckling structures. Phys Rev Lett 99:046101
81. Yin J, Chen X (2010) Elastic buckling of gradient thin films on compliant substrates. Philosophical Magazine Lett 90:423
82. Hutchinson JW (2001) Delamination of compressed films on curved substrates. J Mech Phys Solids 49:1847
83. Priestley JH (1928) The meristematic tissues of the plant. Ann Bot 3:1
84. Considine J, Brown K (1981) Physical aspects of fruit growth. Plant Physiol 68:371
85. Cosgrove DJ (2005) Growth of the plant cell wall. Nat Rev Mol Cell Biol 6:850
86. Dumais J (2007) Can mechanics control pattern formation in plants? Curr Opin Plant Biol 10:58
87. Considine J, Knox RB (1979) Development and histochemistry of the cells, cell walls and cuticle of the dermal system of the fruit of the grape. Vitis vinifera Protoplasma 99:347

Index

X. Chen (ed.), *Mechanical Self-Assembly: Science and Applications*,
DOI 10.1007/978-1-4614-4562-3, © Springer Science+Business Media New York 2013

MIX
Papier aus verantwortungsvollen Quellen
Paper from responsible sources
FSC® C105338

If you have any concerns about our products,
you can contact us on
ProductSafety@springernature.com

In case Publisher is established outside the EU,
the EU authorized representative is:
Springer Nature Customer Service Center GmbH
Europaplatz 3, 69115 Heidelberg, Germany

Printed by Libri Plureos GmbH
in Hamburg, Germany